TRAFFIC AND CONGESTION IN THE ROMAN EMPIRE

'All roads lead to Rome', according to a proverb. A military, commercial and political tool, Roman highways are central to an understanding of Roman civilisation. The Roman Empire is still etched with the remains of Roman roads: lines in the landscape, bridges, dams and tunnels were colonial markings of the Roman world, symbols of the Empire's political and economic success.

As any inhabitant of the modern Western world knows, success brings heavy traffic: civilian transport and goods import and export grew with prosperity, as well as the Roman military, for which the roads were principally built. But did this road system have the capacity to cope with huge traffic flow?

In this book, Cornelis van Tilburg examines the construction of Roman roads in detail and studies the myriad road-users of the Roman Empire: civilians, wagons and animals, the *cursus publicus*, commercial use and the army. In examining the roads, much is revealed of town planning in ancient cities: the narrow paths of older cities, and the wider, chessboard-patterned streets designed to sustain heavy traffic. He discusses toll points and city gates as measures taken to hamper traffic, and concludes with a discussion as to why the local governments' attempts to regulate the traffic flow missed their targets of improving the infrastructure. Traffic was, contrary to modern traffic, a closing entry.

This book will interest any student, scholar or enthusiast of Roman history and culture.

TRAFFIC AND CONGESTION IN THE ROMAN EMPIRE

Cornelis van Tilburg

Routledge
Taylor & Francis Group
LONDON AND NEW YORK

First published 2007 by Routledge
2 Park Square, Milton Park, Abingdon, Oxon OX14 4RN
52 Vanderbilt Avenue, New York, NY 10017

Routledge is an imprint of the Taylor & Francis Group, an informa business

Typeset in Garamond by Laserwords Private Limited, Chennai, India
Printed and bound in Great Britain by TJ International, Ltd, Padstow, Cornwall

British Library Cataloguing in Publication Data
A catalogue record for this book is available
from the British Library

Library of Congress Cataloging-in-Publication Data

Tilburg, C.R. van (Cornelis).
Traffic and congestion in the Roman Empire / C.R. van Tilburg. – 1st ed.
p.cm.
Includes bibliographical references and indexes.
1. Roads–History. 2. Roads–Rome. 3. Roads, Roman I. Title.
TE16.T56 2006
388.10937–dc22
2006016509

ISBN-13: 9780415409995 (hbk)
ISBN-13: 9780415512619 (pbk)
ISBN-13: 9780203968031 (ebk)

CONTENTS

FIGURES

MAPS

TABLES

MAPS

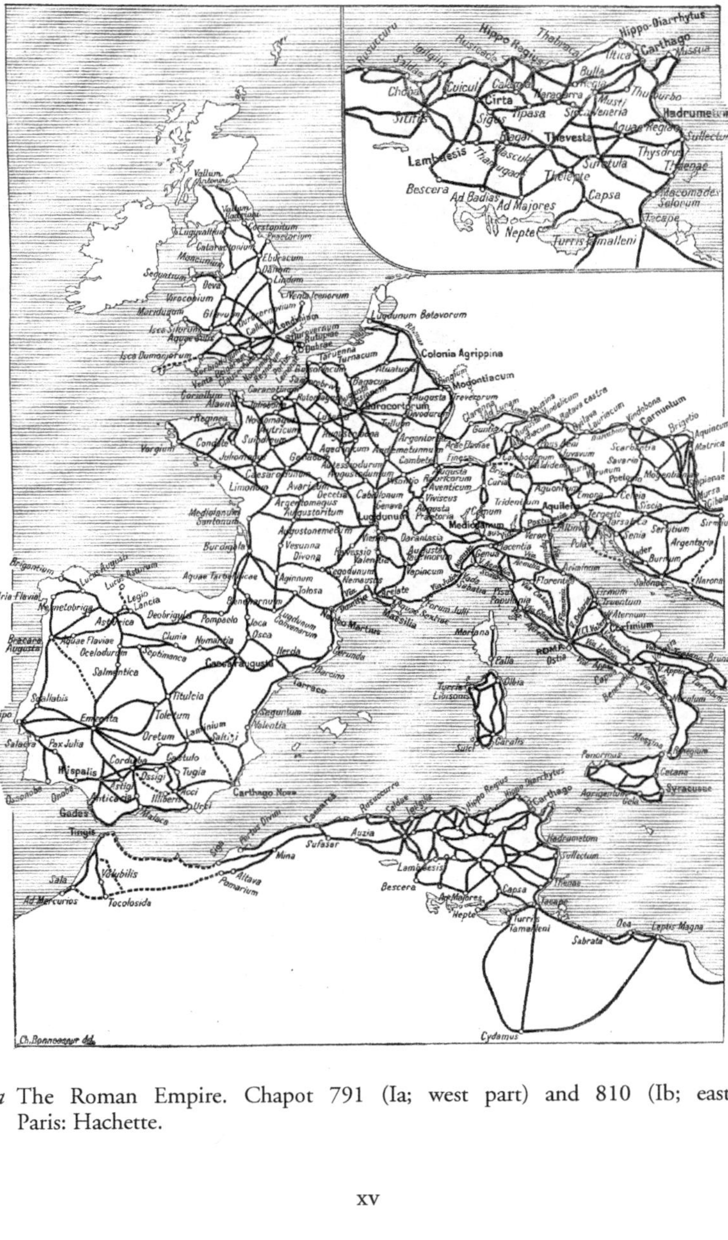

Ia The Roman Empire. Chapot 791 (Ia; west part) and 810 (Ib; east part). Paris: Hachette.

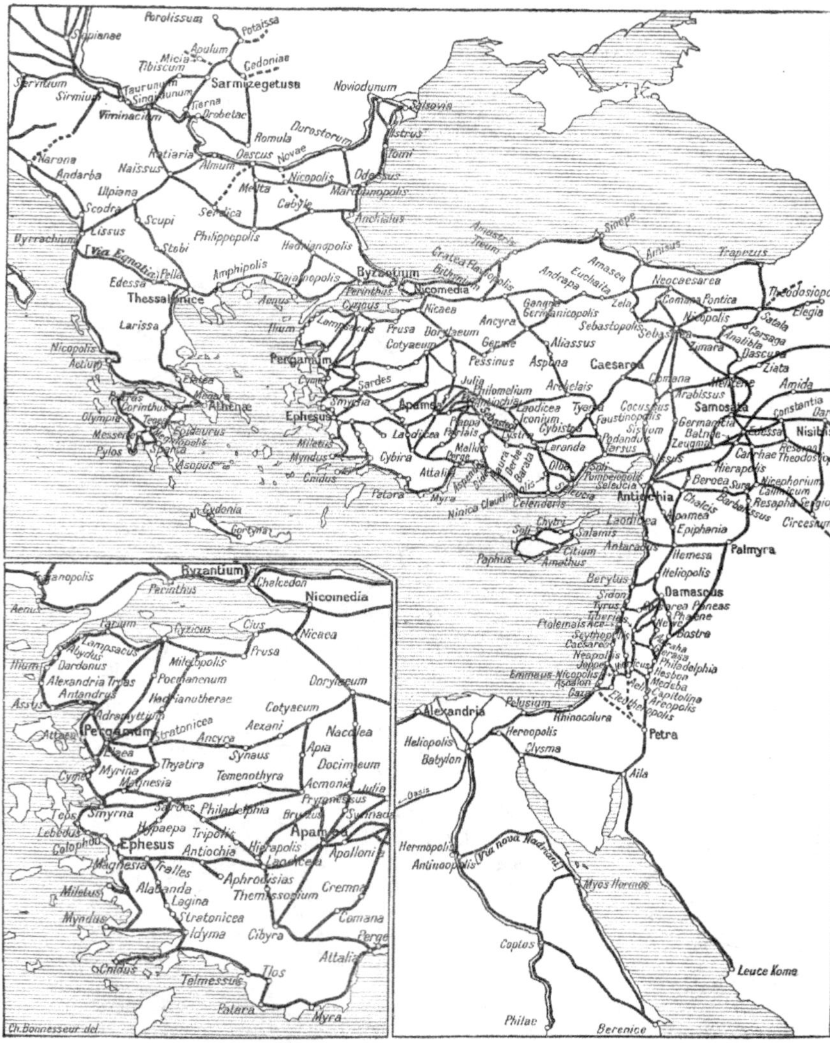

Ib (*continued*)

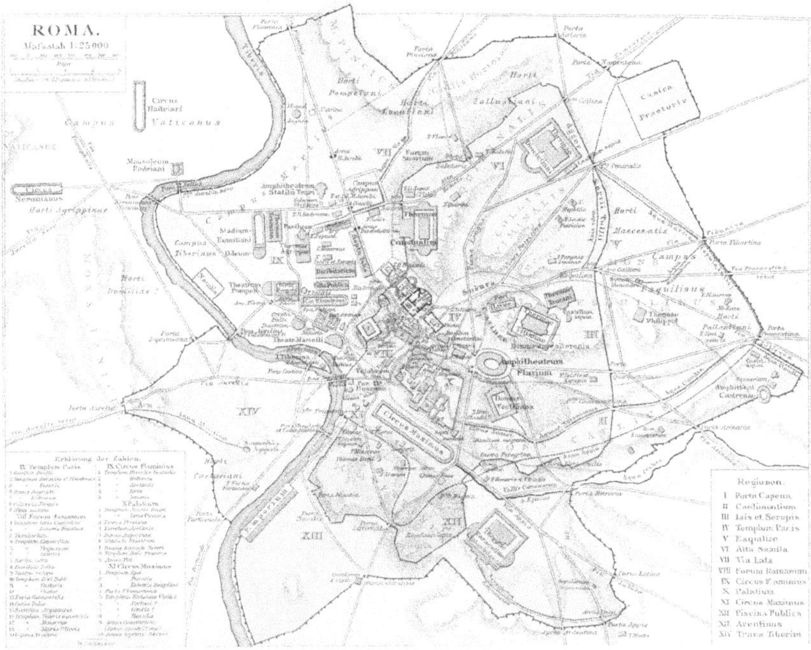

II Rome. G. Droysen, *Historischer Handatlas in sechsundneunzig Karten mit erläuterndem Text.* Bielefeld and Leipzig 1886.

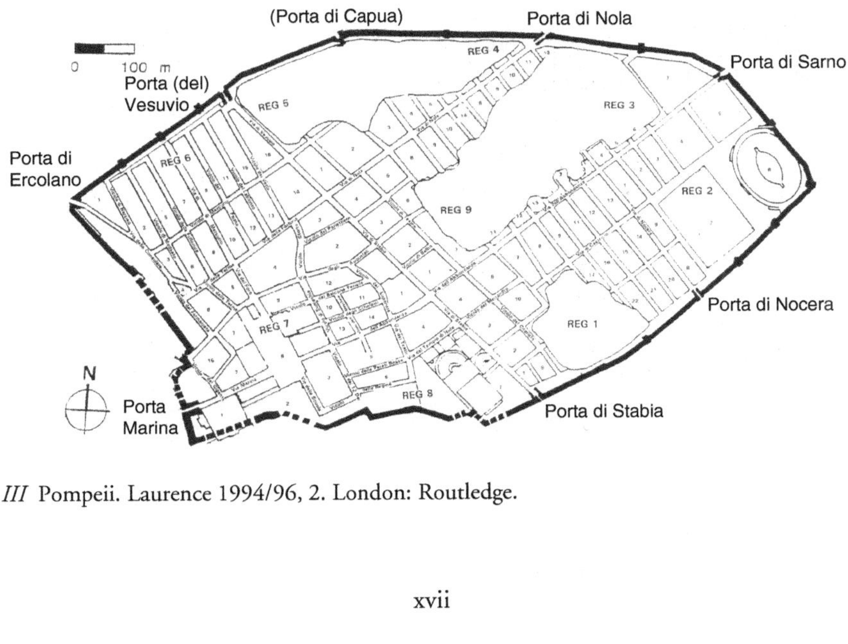

III Pompeii. Laurence 1994/96, 2. London: Routledge.

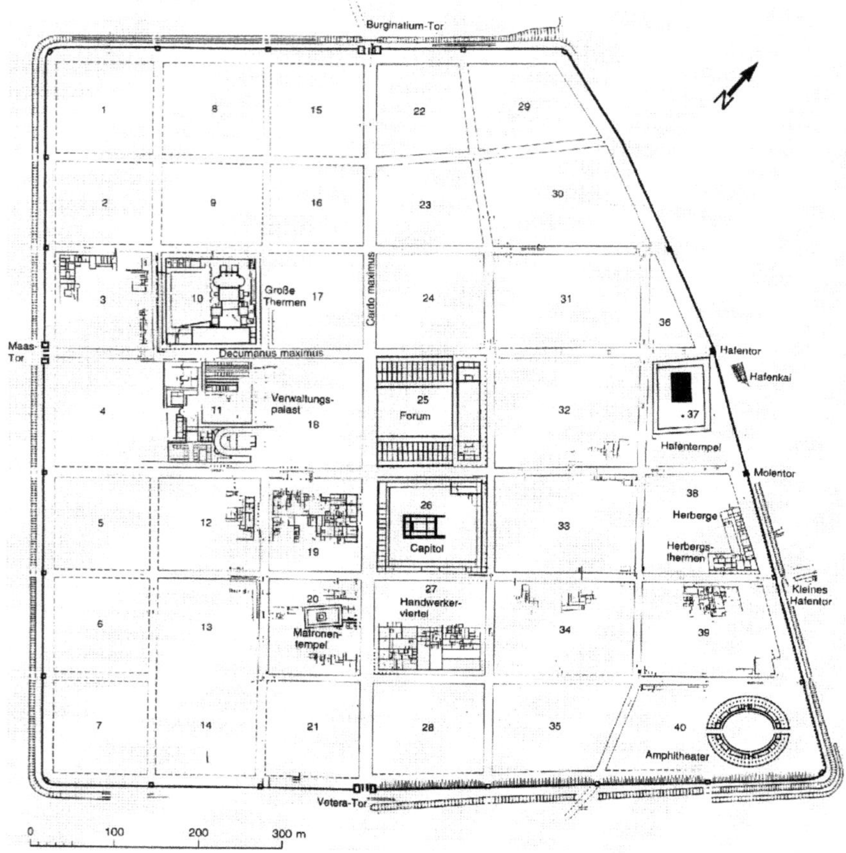

IV Xanten. Heimberg and Rieche 7. H. Stelter, APX.

PREFACE

The idea of researching traffic and traffic policy in the Roman Empire came up when I was waiting in traffic congestion between The Hague and its suburb Zoetermeer. All around me, traffic was completely blocked, although there was a double carriage way with three lanes in each direction, with a railway at each side of the highway. But this extended infrastructure was not able to sustain traffic; it took a long time before cars could move again.

What about this situation in the Roman Empire? In most respects, not only the physiognomy of infrastructure but also the philosophy concerning travelling was quite different. There were no cars, no trains; no highways, no railways, no airlines; traffic was slower, the majority of people travelled on foot, other people on horseback; horses, asses and oxen pulled, in slow motion, wagons for people and goods. They had a lot of time. Sitting in my car, in the jam, my eyes fixed to the rear of the car before me, I saw the image of a simple, rural society without clocks, watches or overflowing diaries, a time with only scarce negative effects of delays.

But was that the truth?

Compared with our busy society – time is money, overflowing diaries take up every minute with sometimes catastrophic results when an appointment is delayed – Roman society, where time was not measured exactly, must have, indeed, been a quiet society. But in the Roman Empire, there were also cities; cities with sometimes more than 100,000 inhabitants, densely built up. Cities with less space for traffic. Indeed, no car traffic; but busy traffic nevertheless. An ass or a horse is not much smaller than a car; a couple of oxen before a loaded wagon constitute a road-user requiring space. No exhaust gases, but the bad smell of excrement of animals in the streets; no motor noise, but shouting, braying, lowing and neighing, completed by the noise of iron wheels on road surfaces.

What role did traffic play in the Roman Empire? An important role, but a different one from nowadays. From the end of the fourth century BC onwards, the Romans organised and realised a road-system to the frontiers of their Empire. This road-system is considered one of their greatest efforts, in Antiquity and later. The roads are paved so well that they are sometimes still in use; in other places, their remains are still visible. The maintenance of this road-system for good communication was essential for the optimal functioning of the Empire as a unity. Even in desolate areas, roads had enough capacity to sustain interurban traffic flow.

Was there talk in the Roman Empire of traffic congestion? There was less traffic than nowadays, but the roads – although, as said, of good quality – did not compare with our modern highways. Traffic congestion is caused by two factors: the capacity (width) of the infrastructure and the volume of the traffic flow. So one can state that in the case of less traffic, but less capacity, traffic congestion can occur nevertheless.

Literary and archaeological sources show that in some places, indeed, there was traffic congestion. The most infamous example of this was Rome itself. The city had, in comparison with other cities, a chaotic and deficient infrastructure, caused by high land prices and unplanned building. A law text, the so-called *Lex Julia Municipalis*, ordered that wheeled traffic by day – with some exceptions – was forbidden inside the city. In other cities, traffic was regulated by means of barricades and one-way traffic; examples of this can be found in Pompeii. The most infamous bottlenecks were formed by the city gates. Especially in older (archaic) cities, where city gates were built in the first instance as part of defence structures and kept as small as possible, traffic congestion must have been a problem. Nevertheless, city gates built in a later period, more suitable for traffic flow, were not always able to sustain it. Excavations in Xanten (Germany) show that (goods) transport could not easily pass one of the harbour gates.

Nowadays, governments are looking for solutions to the growing traffic congestion. Infrastructure is enlarged and improved; new bridges, flyovers and tunnels are constructed and alternative forms of transport like public transport are seen as a (partial) solution. But which answers did the Roman governments give to traffic congestion? The answer is short: as good as nothing. The distance between government and people was large; the local upper class had nothing to do with goods transporters, who had to obey traffic laws; the army and the post company, the *cursus publicus*, had right of way and the man in the street scarcely travelled. We know nothing about complaints of road-users concerning bad accessibility of roads or parts of roads, although there must have been some. The infrastructure – as well as the gates – remained unchanged.

In this book, I discuss the construction of Roman roads in detail and study the myriad road-users of the Roman Empire: civilians, wagons and animals, the *cursus publicus*, commercial use and the army. In examining the roads much is revealed of town planning in ancient cities: the narrow paths of older cities and the wider, chessboard-patterned streets designed to sustain heavy traffic. Toll points and city gates were apparently measures taken to hamper traffic. *Traffic was, contrary to modern traffic, a closing entry.*

ACKNOWLEDGEMENTS

I am greatly indebted to the following people. First I am very grateful to Professor Manfred Horstmanshoff (Universiteit Leiden) for his positive criticism and enthusiasm. When I needed his assistance, he gave it unreservedly; in finding the right connections, he always pointed me in the right direction. Dr Arie van Heck was a great help. He not only gave a large amount of materials concerning city planning and city gates – especially on Rome and Pompeii – but his enthusiasm for the continuing development of the book accelerated its completion.

I gratefully acknowledge that the Faculty of Arts, Universiteit Leiden, particularly the Department of Classics and the Department of History, gave me the time, the infrastructure and the stimulating environment for researching and writing this book. It is impossible to mention the names of all those within these departments who have inspired and advised me.

I am also grateful to Mrs Mariet Samson for her beautiful pictures of the streets in Pompeii. And, last but not least, Jane van Klaveren has given me invaluable support in improving and correcting the English translation of the text.

1

ROADS

The intensity of traffic flow depends on two factors: infrastructure (the road itself) and traffic flow, which makes use of it. In this chapter, I discuss the development, the construction and the maintenance of the infrastructure.

I will look first at the history of the Roman road-system. Starting as a system of primitive unpaved roads (local and regional), it developed into the famous paved road-system because of the growing importance of traffic in the expanding Roman Empire. I will also give attention to the different types of road and their hierarchy.

Later, the construction and width of roads will be dealt with. The width, of course, is the most important factor in sustaining traffic volume. I will also briefly discuss the engineering construction of the roads: bridges, dams and tunnels.

Finally, I will discuss road maintenance. Who was responsible for construction and maintenance of the roads? Epigraphic sources show us a lot of politicians and officials engaging in the financing of roads and road building, as well as landowners and other private individuals.

Development of the Roman road-system

Introduction

The Roman paved road-system, during[1] and after Antiquity, is seen as an enormous achievement and is nowadays one of the best-preserved (archaeological) remains of the Roman Empire. An extended and fine-woven road-system was exceptional before 1700. In prehistoric and medieval Europe, there were no paved road-systems (except the Roman roads, and they had already fallen into ruin by the Middle Ages). Some ancient civilisations such as the Aztec Empire did not even have the facility of paved roads. The roads of the Inca Empire could be compared with the Roman roads, but they did not carry wheeled traffic. The lack of a good infrastructure caused a rapid decline of the Carolingian and Aztec Empires.[2] Over many centuries, the Roman road-system has been thoroughly investigated; remains and traces are found everywhere in the Empire, including the frontiers. The major part of the investigations concerns materials and constructions of the roads; I will discuss this in the next section.

In this section, there are two aspects of the road-system: its history and its development – from prehistoric times to the fifth century AD – and the different types of road, inside and outside the cities and roads in planned landscapes (allocation).

Forerunners of Roman roads

Since the invention of the wheel, probably at the end of the fourth millennium BC in Mesopotamia,[3] people had wanted to use roads which were suitable for wheeled vehicles. The first paved roads were realised, as far as we know, in Mesopotamia; archaeologists have found roads in the neighbourhood of Nineveh (*c.* 2600 BC) and in Babylon asphalt was already used for surfacing roads. However, this material fell into disuse when the Persians captured Babylon.[4]

Persia had a long and important west–east route, the Royal Highway (*βασιλική*), following the course Sardes–Laodicea–Tarsus–Nineveh–Susa–Persepolis, with – including some other branches – a road to Babylon. The roads were not always easily accessible,[5] but had the facility of staging posts and the highway was watched by soldiers.[6]

Minoan Crete also had major roads; there was a north–south connection from Knossos to Gortyn. In the middle of the road, there were flat paving stones, flanked by road metal (Fig. 1.1). The roads were drained and there were watchmen's cabins. In the cities the roads were also paved.[7] From the Mycenaean period a paved road has been found in Troy; this road was built against the slope of a hill.[8] Also in Mycenae itself roads are found with a width of some metres (7–8 feet), suitable for wagons with a width of 4 feet.[9] In the Argolid, some roads had holes to allow water to drain away from the slopes.[10] In Classical Greece there were paved roads leading to holy places, e.g. Eleusis. To avoid the vibrating of statues during processions, roads were supplied with artificial ruts and sleeves, e.g. the Athens–Piraeus and Athens–Sounion roads.[11] Overtaking ruts were also common.[12]

The Greeks had several different terms for 'road': the most used and best known is *ὁδός*; a road suitable for wagons was called *ἁμαξιτός*. Further, they had words for the less important roads and paths.[13] However, the Greeks did not have an extended road-system, because of the rocky landscape and the political rivalry of city-states. In these circumstances road building was not only difficult, but also impractical. Strabo says that the Greek roads were of bad quality and badly drained.[14] Also Pausanias mentions rough roads. Livy, however, describes a well-built road in Macedonia, with bridges and paved surfaces.[15]

Figure 1.1 The oldest paved road of Europe, Knossos, Crete. Photo: C.R. van Tilburg.

In the cities, the streets were usually unpaved. Aristophanes mentions muddy streets in Athens.[16] In 320/319 BC, the ἀγορανόμοι (street authorities) of Athens repaired the roads to Piraeus and the sanctuaries of Zeus Soter and Dionysos, amongst others, by means of a bridge over the Kephisos river.[17] In the fourth century BC, in Thebes, the function of τελέαρχος was created; he had to maintain the condition of the streets.[18]

Alexander the Great ordered a road to be built by the Thracians,[19] and his successors, the Hellenistic kings, completed new roads, following the concept of the Persian Royal Highway.[20]

Main streets in the Hellenistic cities were wide and well paved. The most important roads, λεωφόροι, were twice as wide as other streets.[21] Strabo mentions wide streets with a paved surface (λιθόστρωτοι) in Smyrna;[22] Alexandria, Seleucia, Ephesus and Gerasia had the convenience of streets with colonnaded sidewalks[23] and Antioch's main street was even paved with marble.[24]

In these cities, the streets had good and efficient drainage and the local governments maintained the draining of waste water out of the sewers. But the most important reason for keeping the streets clean was to present a good appearance rather than to fight diseases and epidemics. The relationship between good health and drainage had not yet been discovered.[25]

The start of Roman road building

Servius and Isidorus tell us that the Romans adopted the technique of road building from the Carthaginians.[26] There were small streets and roads in Carthage,[27] but nowadays most scholars state that the roots of Roman road building are to be found in the Etruscan (or Greek) cities in south Italy.[28] From these Greek cities the Romans adopted the use of mortar as cement.[29]

The oldest remains of roads in Italy are from the periods before the Etruscans: the Bronze Age and the Terramare period (second millennium BC and the beginning of the first millennium BC). Little is known about these roads; they were probably unpaved tracks.[30] The Etruscans used drainage canals beside the roads.[31] In Faesulae, Perugia, Saturnia and Graviscae Roman roads had Etruscan forerunners with wheel ruts, built of lime mortar.[32] Quilici gives some examples of archaic paved roads in central Italy, with a width of some 3 m;[33] this width satisfies the so-called 8 feet-norm, which will be discussed later. The city-state society of the Etruscans prohibited the realisation of an extended road-system, as in Greece.

The streets of the Etruscan cities were paved and well drained. In the city of Marzabotto (fifth century BC) four main streets are found with a width of 15 m, including sidewalks and stepping stones as in Pompeii.[34] The well-preserved Via degli Inferi in Cerveteri was flanked by graves and wheel ruts.[35]

The first planned Roman road: Via Appia

If we conclude that (partially paved) roads and streets were known in Italy, we can also assume that Rome had its roads from the beginning – for the supply of food and materials for the newly founded city – but these roads were not planned and (probably) not paved. The Italian archaeologist Cozzo assumes that there was an Etruscan road from Tiber Island to the Forum Romanum.[36] Albeit unpaved, the first exit routes of royal and Republican Rome were suitable for vehicles; Livy describes the flight of the Vestal Virgins, where Lucius Albinius had at his disposal a wagon.[37] Livy also mentions Via Gabina (a forerunner of Via Praenestina) in 208 AUC (545 BC),[38] Via Latina in 266 AUC (487 BC)[39] and Via Salaria in 393 AUC (360 BC).[40] These roads did not yet have the names of officials, but more usually of places.

In general one can say that the systematic construction of the Roman road-system makes its entry in 312 BC,[41] when Appius Claudius Caecus, during his censorship (312-308 BC), gave the order to build Via Appia from Rome to Capua; it was also the first road to bear the name of the politician responsible.[42] Some scholars, however, disagree with this statement. Radke states that a censor did not have permission to construct a road outside of the *ager publicus* and that the construction of Via Appia took place during a later period, in the years 307, 296 and 295 BC.[43] According to Pekáry, censors did not engage with roads, and systematic road planning did not take place before the second century BC.[44]

Livy mentions two finished parts of Via Appia after 312: the paved route from Porta Capena to the Temple of Mars, close to the first milestone in 296 BC[45] and the extension to Bovillae in 293 BC.[46] Also according to Livy, the censors Q. Fulvius Flaccus and A. Postumius Albinus paved the streets within Rome with stones and outside the city with gravel.[47]

The most important reason for completing Via Appia as a paved road was a political and military one: the Romans were defeated in 315 at Lautulae (between Rome and Capua) and in 314 a rebellion in Capua followed.[48] Capua was hostile to Rome; in the Second Punic War the city was on the side of Hannibal.

The route of Via Appia followed and replaced an older coast route[49] and shows the typical straight route of the later Roman roads.[50] The planning is probably influenced by the Pythagorean ideas of the Greeks of south Italy;[51] Appius was an admirer of Hellenism and of the ideas of Pythagoras.[52] From Forum Appi to Feronia, the route passed the Pontine Marshes, so it was necessary to dig a canal along the road. Until *c.* 100 AD gravel was used here; under Nerva or Trajan gravel was replaced by stone.[53]

As mentioned, for a long time Capua was the terminus of Via Appia.[54] Later, the road was extended first to Venusia, then to Tarente and, finally, to Brindisi.

Later roads

After the start of the construction of Via Appia, the Romans soon constructed other roads.[55] Via Clodia, going north to Etruria, also dates from the end of the fourth century BC;[56] the construction of Via Valeria to Alba Fucens, probably started in 304 BC;[57] Via Aurelia, running north-west, following the coastline of Etruria, dated from 241 BC;[58] Via Cassia, date unknown, followed its course to the north, also to Etruria[59] and Via Flaminia (from 220 BC)[60] also followed its course to the north to Rimini on the Adriatic Sea.[61] Via Aemilia (from 187 BC)[62] followed the Adriatic coastline to Piacenza.[63]

The first road outside Italy was Via Egnatia, from Durrës to Thracia. Polybius mentions this road for the first time, mid second century BC.[64] Via Domitia, from Ampurias to the Rhône, was realised from 121 BC and derives its name from the proconsul of Gallia, Cn. Domitius Ahenobarbus. Cicero speaks about this road in his *Pro Fonteio*.[65] C. Gracchus also took care of road building in 123 BC; he stimulated and propagated the extension of road building in his proposed law *Lex Sempronia viaria*.[66]

In Caesar's time, everywhere in the Empire paved roads were in good condition,[67] but during the Civil Wars there were hardly any new roads built.[68] After this period, a boom in road construction came and Augustus and his successors repaired the older, existing roads on a large scale. Not all these new roads got names, and in the provinces, where the majority of new imperial roads were constructed, names were rare.

During the Empire, the Roman Empire had 80,000–100,000 km of paved roads. Forbes gives a total length of 290,000 km roads; 86,000 km were main roads.[69] The common image of Roman roads – long, straight paved roads like Via Appia near Rome – is only partially correct. The major part of the road-system was simpler and in the provinces gravel roads were common. Even parts of Via Appia were unpaved until the second century AD,[70] and some sections of the roads were too small for passing or overtaking.[71]

The road-system was in use for a long time. When more and more armies used the roads in the third and fourth centuries AD, the maintenance of the roads was a matter of great importance. From the beginning of the fifth century onwards, the road-system started to decline. During the reign of Honorius and Arcadius the condition of many roads was bad.[72] Claudius Claudianus describes Via Flaminia as dusty (*pulverulenta*),[73] which most likely refers to bad maintenance, and the poet Rutilius Namatianus gives a description of how in 417, seven years after the sack of Rome, roads and bridges were destroyed.[74] Other roads were still intact; a hundred years later Procopius writes that the Byzantine general Belisarius used Via Latina in his campaign against the Ostrogoths;[75] Via Appia was then also still in good condition.[76] But after that period, the declining condition of the road-system continued. In the Germanic kingdoms there was no place for Roman roads and their maintenance. Over the centuries, most of the roads disappeared. Sometimes their traces survived as straight lines, but even the knowledge of their Roman origin had disappeared and people ascribed them to Charlemagne, Queen Brunehildis or the Devil.[77] Sometimes roads remained in use in the Middle Ages, including in Britain, but they were badly maintained for traffic, which had reduced with the declining population.[78] New roads were no more than traces and the 'right of driving'.[79]

The development of streets in Rome

In 174 BC the censors Q. Fulvius Flaccus and A. Postumius Albinus gave the order that roads in Rome had to be paved.[80] But this did not mean that before that time there were no paved roads in Rome; the first paved road of Rome was Clivus Publicius in 238 BC,[81] running from the Forum Boarium to Porta Raudusculana. Before 174 BC, some parts of the Forum Romanum were also paved.[82]

Rome had a chaotic street-system; Livy ascribes it to the hasty reconstruction of the city after the invasion of the Gauls in 390 BC.[83] Later in the fourth century BC the Servian Wall was erected, including a number of gates, between thirteen and twenty, and Rome had as many exit routes.[84]

Livy also mentions the paving of Clivus Capitolinus.[85] Outside Rome he speaks about the paving of a road in Pesaro (on the Adriatic Sea, south-east of Rimini).[86] The German scholar Mommsen states that in 174 BC not all roads were completely paved; confirming that the censors had not ordered them to be so. He makes a distinction between two meanings of the word *via*; a road in the common sense and a roadway for drivers. Maybe the sidewalks were paved, but not the roadways; the roads on slopes, the so-called *clivi*, were explicitly mentioned.[87] Mommsen also states that existing roads and streets, inside and outside Rome, were maintained.[88] In Pompeii, in the fourth century BC, the paving of streets had already started, and in the second century sidewalks were constructed.[89]

Unfortunately, we are not very well informed about later street projects in Rome. Sulla gave orders to pave parts of the Forum Romanum[90] and Caesar used marble and travertine for the pavements of the Forum Romanum and Via Sacra.[91] We have to assume that after the middle of the first century BC the majority of the streets, including sidewalks, and *fora* are paved.[92]

Different types of roads and streets

There are various ways of classifying different types of roads. In the Roman Empire, there were a lot of different names for roads and streets, classified by physical and juridical aspects and traffic purposes; not every form of traffic was permitted or able to use a certain road or street. I give the following list:

- *Via*: the most usual word used for 'road'. A *via* can be inside or outside a settlement or city and has a minimum width of 8 feet (about 2.40 m) or more;[93] there are narrower and wider *viae*. People can drive or walk on *viae*.[94] There are different types of *viae*. Briefly the following most important distinctions can be made:[95]

  - *via consularis* – a road built by a consul[96]
  - *via militaris* – a road built by or for an army[97]
  - *via plostralis* – a road suitable for wagons (*plostra or plaustra*)[98]
  - *via praetoria* – a road built by a praetor[99]
  - *via privata* – a private road, built by a private person or local authority[100]
  - *via publica* – a public road[101]
  - *via vicinalis* – a local road, connecting to a highway or two highways.[102]
- *Actus*: a road with a width of 4 feet, accessible for pedestrians, vehicles and animals, without passing or overtaking each other.[103]
- *Iter*: a road or path with a width of 2 feet, only accessible for pedestrians, horsemen and sedans;[104] also, in more common parlance, a street or road in the city or countryside.[105]
- *Ambitus*: a circular road, circling (mostly) a building.[106] The Law of the Twelve Tables, discussing defence structures, talks of *ambitus parietis*.[107]
- *Semita*: 'half road', in the countryside: a very narrow path. In the city: a small lane (e.g. the Alta Semita in Rome) or sidewalk.[108]
- *Callis*: a path in the mountains or woods. These paths were used in particular for the transportation and relocation of cattle.[109]
- *Trames*: a branch road in the countryside.[110]
- *Deverticulum, diverticulum*: branch road or street.[111]
- *Clivus*: street on a slope.[112]
- *Vicus*: narrow or wide street with houses in a settlement. Rome was divided into *regiones*, and they in turn into *vici*, streets including intersecting streets; together they form a district.[113]
- *Angiportum, angiportus*: small street, mostly nameless.[114]
- *Fundula*: cul-de-sacs.[115]
- *Scalae*: ascending street with steps.[116]
- *Platea*: wide street. Plautus uses the Greek loanword *platea* (πλατεῖα), but the word is used first in Late Antiquity in common parlance. Isidorus gives the meaning as 'wide street', but in later times it took on the significance of 'place'.[117] A *platea* could allow a wagon (*carpentum*) driving over a fifth part of it (*quintana*).[118] According to Homo, in Rome only Via Lata and Via Nova were *plateae*.[119]
- *Forum*: square. Public squares are scarce in Roman cities; planned cities like *coloniae* had usually only one central place. A *forum* could also have the function of market-place (e.g. Forum Boarium). In Rome there were, besides the *fora*, also the *areae*, the *campi* and the *compita*. *Areae* were the most common of them and could have different dimensions; the biggest *area* was situated in front of the Baths of Titus, with dimensions of 83 by 12 m. *Campi* were scarcer and had the functions of places for

walking. *Compita*, in fact, were crossings.[120] The *forum* was typically Roman; Hellenistic cities, although also planned with a Hippodamic street-system, did not have such a central square.

Juridical aspects

Besides the terms for roads and streets concerning their physical character, there are juridical significances. As said, *via* is the most common definition of a road; it is a road accessible for all types of traffic – coaches, vehicles, pedestrians and animals – also in a legal context: it was not only *possible* to make use of it, but also *permitted. Via* is a combination of two other juridical terms: *actus* (*ius agendi*) and *iter* (*ius eundi*). This classification is composed by the jurist Ulpianus.[121]

Actus is the right to use a road in a vehicle and to drive a herd of animals.[122] As mentioned, *actus* is also a road with a width of 4 feet, but this interpretation is rarely used.[123]

Iter is the right to use a road on horseback, on foot or by sedan.[124] In fact, any other function is not possible, because a road (path) with the status of *iter* is too small (2 feet) for a vehicle.[125] One should say that the right of *actus* includes the right of *iter*, because *actus* has a higher rank than *iter*. Ulpianus makes an exception, however: the so-called right *actus sine itinere*: one can only make use of a road in certain cases to go somewhere, the 'right of way'.[126]

Another classification of Ulpianus is formed by *viae publicae, viae privatae* and *viae vicinales. Viae publicae* were built on public land and accessible to everyone. *Viae vicinales* were roads connecting *viae publicae* with settlements, villages and each other. These roads were also paved roads and constructed by a central or more commonly a local authority.[127] Finally, *viae privatae* were constructed mostly by landowners to access their land. These roads were inaccessible to unauthorised people.[128]

Roads in geodesy

In geodesy roads have specific names which I will now discuss.

As in the case of planning of new towns (*coloniae*), as well as in the case of allocation, the basis was a chessboard pattern.[129] The north–south roads were called *cardines* (singular: *cardo*);[130] the main road was called the *cardo maximus*. The east–west roads were called the *decumani* (or *decimani*)[131] and, of course, the main road the *decumanus maximus*.[132] We also meet the words *cardo* and *decumanus*, besides geodesy, in the central axes of planned Roman towns.

From the *cardo maximus* and *decumanus maximus* onwards, the land was divided into square or rectangular plots; a road or path, called *limes*, surrounded each plot.[133] Every fifth *limes*, numbered from the *cardo* or

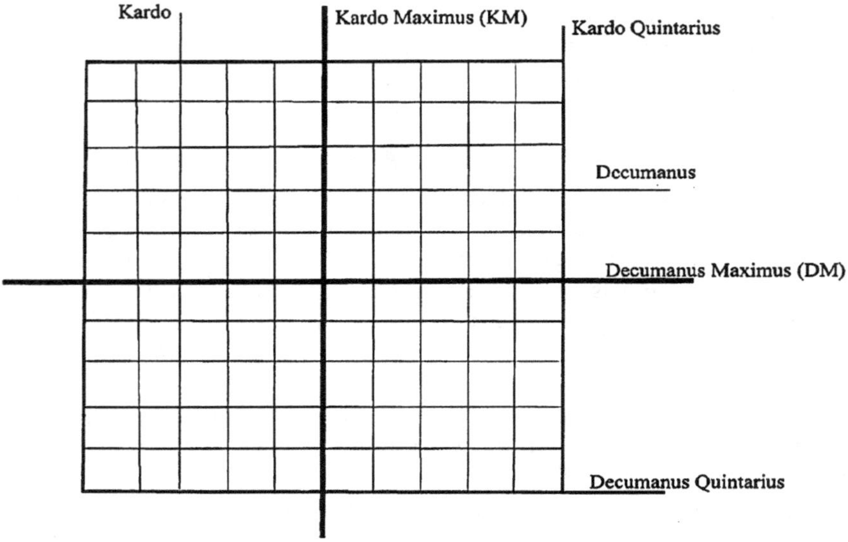

Figure 1.2 Cardo maximus, decumanus maximus and *quintarii.* Davies 68. Stroud: Tempus Publishing.

decumanus was called *limes quintarius* or *actuarius* (see Fig. 1.2);[134] a *limes quintarius* was wider than the other *limites* (20 feet against 12 or 8 feet).[135] The other *limites* were called *subruncivi*; these were the narrowest paths with the prescribed width of 8 feet.[136] I will discuss the width of these roads in the next section.

Summary and conclusion

The development of a road-system did not take place everywhere in the same way; it depended on the physical landscape and political structure of the region where roads were realised. One can say that in states with a central government, like Persia, the road-system proliferated the most; in such states, we find paved roads with a length of hundreds or even thousands of kilometres. In societies where people lived in city-states, like Classical Greece, the road-system was not only shorter, but also less developed and worse paved, the result of rivalry among city-states. Here, there was no need to realise a coherent road-system.

The Roman Empire had undergone a development from city-state to Empire; the road building was dependent on this development. From the fourth century BC onwards, when the Romans began the conquest of Italy, the Romans built a paved and straight road-system, inside and outside their settlements, villages and cities, enabling fast army movements and stimulating trade. For building purposes they used a combination of techniques and sciences drawn from other people: Persians, Etruscans, Greeks and possibly Carthaginians. In newly conquered areas the roads which had ended at former frontiers were extended to accomodate new frontiers. Finally, the Romans had a road-system with a total length of 300,000 km in the whole Empire, including a main road-system with a length of 80,000–100,000 km.

The construction of long-distance paved roads started earlier than that of local paved roads; the reason is that the Romans, in the first instance, built their roads for the army.

The Romans recognised different types of roads and streets, categorised by physical and juridical aspects. The names of the physical aspects are like our modern 'street', 'alley', 'lane', etc. The names of the juridical aspects are like our 'cycle path', 'foot path', etc.

The glory of the Roman road-system

I now present some quotations from classical authors who praised the Roman road building. Firstly a citation from Plutarch:

> Ἐσπούδασε δὲ μάλιστα περὶ τὴν ὁδοποιΐαν, τῆς τε χρείας ἅμα καὶ τοῦ πρὸς χάριν καὶ κάλλος ἐπιμεληθείς. εὐθεῖαι γὰρ ἤγοντο διὰ τῶν χωρίων ἀτρεμεῖς, καὶ τὸ μὲν ἐστόρνυτο πέτρᾳ ξεστῇ, τὸ δὲ ἄμμου χώμασι νακτῆς ἐπυκνοῦτο. πιμπλαμένων δὲ τῶν κοίλων καὶ ζευγνυμένων γεφύραις ὅσα χείμαρροι διέκοπτον ἢ φάραγγες, ὕψος τε τῶν ἑκατέρωθεν ἴσον καὶ παράλληλον λαμβανόντων, ὁμαλὴν καὶ καλὴν ὄψιν εἶχε δι' ὅλου τὸ ἔργον.
>
> But he (C. Gracchus) busied himself most earnestly with the construction of roads, laying stress upon utility, as well as upon that which conduced to grace and beauty. For his roads were carried straight through the country without deviation, and had pavements of quarried stone, and substructures of tight-rammed masses of sand. Depressions were filled up, all intersecting torrents or ravines were bridged over, and both sides of the roads were of equal and corresponding height, so that the work had everywhere an even and beautiful appearance.
>
> (Plu. *CG* 7.1)

The third-century church father Tertullian saw the roads as bringers of civilisation:

> *Certe quidem ipse orbis in promptu est cultior de die et instructior pristino. Omnia iam pervia, omnia nota, omnia negotiosa, solitudines famosas retro fundi amoenissimi oblitteraverunt, silvas arva domuerunt,*

> *feras pecora fugaverunt, harenae seruntur, saxa panguntur, paludes eliquantur, tantae urbes quantae non casae quondam. Iam nec insulae horrent nec scopuli terrent; ubique domus, ubique populus, ubique respublica, ubique vita.*
>
> The world is known better each day, more cultivated and civilized than the previous day. Everywhere roads are paved, every region is known, every country opened for trade. Beautiful estates replaced former infamous deserts, there are fields where once have been woods, herds of cattle have wild animals driven away; sand is sown, rocks are split, marshes drained. There are many cities where are in former times no houses at all. Islands do not shiver people, rocks make them not frightened. Everywhere are houses, everywhere human habitation, everywhere governments, everywhere life.
>
> (Tert. *De anima* 30.3, tr. C.R. van Tilburg)

The sixth-century writer Procopius praises the Via Appia, which at that time was 800 years old:

> Ἔστι δὲ ἡ Ἀππία ὁδὸς ἡμερῶν πέντε εὐζώνῳ ἀνδρί ἐκ Ῥώμης γάρ ἄυτη ἐς Καπύην διήκει. εὖρος δέ ἐστι τῆς ὁδοῦ ταύτης ὅσον ἁμάξας δύο ἀντίας ἰέναι ἀλλήλαις, καὶ ἔστιν ἀξιοθέατος πάντων μάλιστα. τὸν γὰρ λίθον ἅπαντα, μυλίτην τε ὄντα καὶ φύσει σκληρόν, ἐκ χώρας ἄλλης μακρὰν οὔσης τεμὼν Ἄππιος ἐνταῦθα ἐκόμισε· ταύτης γὰρ δὴ τῆς γῆς οὐδαμῆ πέφυκε. λείους δὲ τοὺς λίθους καὶ ὁμαλοὺς ἐργασάμενος, ἐγγωνίους τε τῇ ἐντομῇ πεποιημένος, ἐς ἀλλήλους ξυνέδησεν, οὔτε χάλικα ἐντὸς οὔτε τι ἄλλο ἐμβεβλημένος. οἱ δέ ἀλλήλοις οὕτω τε ἀσφαλῶς συνδέδενται καὶ μεμύκασιν, ὥστε ὅτι δὴ οὐκ εἰσὶν ἡρμοσμένοι, ἀλλ᾽ ἐμπεφύκασιν ἀλλήλοις, δόξαν τοῖς ὁρῶσι παρέχονται. καὶ χρόνου τριβέντος συχνοῦ δὴ οὕτως ἁμάξαις τε πολλαῖς καὶ ζῴοις ἅπασι διαβατοὶ γινόμενοι ἐς ἡμέραν ἑκάστην οὔτε τῆς ἁρμονίας παντάπασι διακέκρινται οὔτε τινὶ αὐτῶν διαφθαρῆναι ἢ μείονι γίνεσθαι ξυνέπεσεν, οὐ μὴν οὐδὲ τῆς ἀμαρυγῆς τι ἀποβαλέσθαι. τὰ μὲν οὖν τῆς Ἀππίας ὁδοῦ τοιαῦτά ἐστι.
>
> Now the Appian Way is in length a journey of five days for an unencumbered traveller; for it extends from Rome to Capua. And the breadth of this road is such that two wagons going in opposite directions can pass one another, and it is one of the noteworthy sights of the world. For all the stone, which is mill-stone and hard by nature, Appius quarried in another place far away and brought there; for it is not found anywhere in this district. And after working

> these stones until they were smooth and flat, and cutting them to a polygonal shape, he fastened them together without putting concrete or anything else between them. And they were fastened together so securely and the joints were so firmly closed, that they give the appearance, when one looks at them, not of being fitted together, but of having grown together. And after the passage of so long a time, and after being traversed by many wagons and all kinds of animals every day, they have neither separated at all at the joints, nor has any one of the stones been worn out or reduced in thickness, – nay, they have not even lost any of their polish. Such, then, is the Appian Way.
>
> (Procop. *Goth.* 5.14.6-11)

These citations mainly praise the functioning of the roads. Finally, a passage of a poem of Statius (45–96 AD) about the building of the Via Domitiana, a branch of the Via Appia, to Naples. It is one of his poems from his work *Silvae* and is not only a eulogy of the beautiful landscape, the pleasure of using the road and Domitian as a person, but also in praise of the road workers.

Via Domitiana

Hic segnis populi vias gravatus
et campos iter omne detinentes
longos eximit ambitus novoque
iniectu solidat graves harenas
gaudens Euboicae domum Sibyllae
Gauranosque sinus et aestuantes
septem montibus admovere Baias.
Hic quondam piger axe vectus uno
nutabat cruce pendula viator
sorbebatque rotas maligna tellus,
et plebs in mediis Latina campis
horrebat mala navigationis;
nec cursus agiles, sed impeditum
tardabant iter orbitae tacentes,
dum pondus nimium querens sub alta
repit languida quadrupes statera.
at nunc, quae solidum diem terebat,
horarum via facta vix duarum.
non tensae volucrum per astra pennae
nec velocius ibitis, carinae.
    Hic primus labor incohare sulcos
et rescindere limites et alto
egestu penitus cavare terras;
mox haustas aliter replere fossas

et summo gremium parare dorso,
ne nutent sola, ne maligna sedes
det pressis dubium cubile saxis;
tunc umbonibus hinc et hinc coactis
et crebris iter alligare gomphis.
o quantae pariter manus laborant!
hi caedunt nemus exuuntque montes,
hi ferro scopulos trabesque levant;
illi saxa ligant opusque texunt
cocto pulvere sordidoque tofo;
hi siccant bibulas manu lacunas
et longe fluvios agunt minores.
hae possent et Athon cavare dextrae
et maestum pelagus gementis Helles
intercludere ponte non natanti.
his parvus, nisi di via<m>[137] *vetarent,*
Inous freta miscuisset Isthmos.

'tis he who, brooking ill the slow journeys of his people and the plains that clog every minute of the road, sweeps away tedious windings and lays a new solid paving upon the weary sands, rejoicing to bring the Euboian Sibyl's home and the dells of Gaurus and sweltering Baiae nearer to the seven hills.

Here on a time the tardy traveller, borne on a single axle, was balanced on the swaying pole, while the unkindly earth sucked in the wheels, and Latin folk shuddered in mid-plain at the evils of a sea-voyage; nor could carriages run nimbly, but the noiseless track made their course hampered and slow, while the fainting beast, complaining of a too heavy load, crept on beneath its lofty yoke. But now a journey that once wore out a solid day is performed in scarce two hours. No swifter fare ye through the heavens, ye birds with outstretched pinions, nor will ye more swiftly sail, ye ships.

The first labour was to prepare furrows and mark out the borders of the road, and to hollow out the ground with deep excavation; then to fill up the dug trench with other material, and to make ready a base for the road's arched ridge, lest the soil give way and a treacherous bed provide a doubtful resting-place for the o'erburdened stones; then to bind it with blocks set close on either side and frequent wedges. Oh! how many gangs are at work together! Some cut down the forest and strip the mountain-sides, some plane down beams and boulders with iron; others bind the stones together, and interweave the work with baked sand and dirty tufa; others by dint of toil dry up the thirsty pools, and lead far away the lesser streams. These hands could hollow out Athos, and bar with no floating bridge the

> doleful sea of moaning Helle. These hands, did not the gods forbid the passage, had made Ino's puny Isthmus mingle the sundered seas.
>
> (Stat. 4.3, 20–60)

Construction and width of Roman roads

Introduction

We do not have much information on road construction in Antiquity. Our knowledge of Roman road building is completely based on archaeological and epigraphical sources, but these are available in a vast amount. Parts of roads are still intact, like Via Appia immediately outside Rome. In other places, the original road surface has vanished, but the course of the roads is still visible in the landscape in the form of a straight line. In some other places, the course is not visible at all, but in the case of excavations traces of the former course are found. We can conclude that road construction and building was a complicated matter, because the Romans aimed to build roads, as much as possible, in a straight line (*linea recta*); natural obstacles like mountains, rivers and swamps had to be crossed.

Because the Roman Empire had many types of landscape, road building had to be moulded to fit. Via Appia passing the Pontine Marshes is discussed above. For crossing swamps and mountains the Romans invented various solutions. I will discuss briefly the construction of bridges, dams and tunnels.

A crucial point was the width of roads and streets; there were various widths for the different types. On the interurban scale, roads had enough space to sustain the flow of traffic, as did the thoroughfares inside the cities. Side streets, however, were usually much narrower. This theme will be discussed below, pp. 26–31.

Paved and unpaved roads

In the previous section two classifications of roads were discussed, categorised by physical and juridical aspects. A third classification can be made in a technical sense: the distinction between unpaved roads (*viae terrenae*) and paved roads (*viae munitae*[138] or *viae stratae*). Paved roads in their turn can be distinguished as *viae silice stratae* (roads paved with stones) or *viae glarea stratae* (roads paved with gravel).[139]

Standard construction of roads

To construct a road in a terrain with a standard soil – no rocks, no desert, no swamp – one first dug out a bed for the road mass (*agger*). In the stratification of the road mass there are, roughly, four layers: from the lower to the upper surface one can distinguish the *statumen, rudus, nucleus* and *pavimentum* or

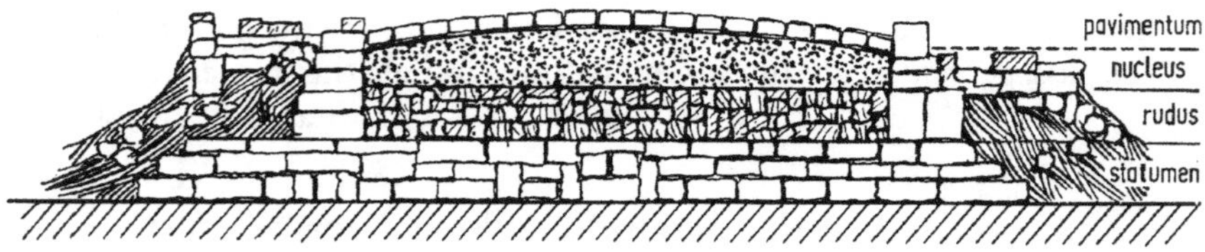

Figure 1.3 Cross-section of a standard Roman road. Radke 1903–1978, 1439–1440. Stuttgart: Metzler.

summum dorsum.[140] The *statumen* is a layer of stones laid in mortar, with a thickness of 20–30 cm. The following layer called *rudus* is like the *statumen* layer, but contains fewer stones and more mortar. The thickness of the *rudus* is 30–50 cm. *Statumen* and *rudus* form, in fact, the foundation of the road. In absorbing groundwater, the mortar will harden. This hardening function also encompasses the following layer, the *nucleus.* This layer has even more mortar and fewer stones and has a thickness of about 30 cm. The upper layer, *pavimentum* or *summum dorsum,* is the road surface, paved with stones or gravel, and has a thickness of 20–40 cm. In total, the thickness of the road mass, including the road surface, is 1 to 1.40 m.[141] During times of repair and with construction of new layers the thickness can increase to 6 m.[142]

To discharge rainwater, roads usually had drainage canals. In the case of a slope, one side of the road could have drainage canals or ruts on the upper side.[143]

The most comfortable roads of the Empire were gravel roads; their surface was relatively flat. This type of road was the most common in the whole Empire.[144] It was called *via glarea strata,* and was very suitable for thoroughfares; in cities, however, where traffic was more intensive and speed was lower, one would prefer a *via silice strata.* As we see, inside the walls of Herdonia, situated along Via Trajana (south-east Italy), there are *viae silice stratae*; beyond the north-east gate, the road surface changes to a *via glarea strata,* where gravel is laid on a former road surface of paving stones.[145] *Viae silice stratae* were a common phenomenon in very densely populated regions in Italy.[146]

Figure 1.3 shows the cross-section of a standard Roman road. Besides that, Forbes and Ortalli show other cross-sections of local Roman roads.[147] Generally one can say that, in a road mass stratigraphy, as you go from the bottom to the top there is more mortar and fewer stones.[148]

The construction of mountain roads

The construction of mountain roads caused more problems. The road constructors faced the difficulty not only of steep slopes, but also of rocky ground, difficult to dig. As a result of landslides, severe snowfall (in winter many passes were full of snow) and avalanches the roads were often blocked. Little rivers suddenly became wide and heavy torrents which in springtime could demolish parts of roads.[149] It is clear that road builders constructed the roads in the river valleys as much as possible.[150] They also tried to build the roads on the sunny side of the mountain slopes to diminish the danger of avalanches.[151]

The roads crossing the Alps could not be completed before 15 BC, when Tiberius and Drusus successfully fought mountain tribes.[152] Drusus started the construction of Via Claudia Augusta, from Trento to Augsburg; the work was finished by Claudius.[153] Troubles on the Danube frontier in the second century AD made necessary the construction of a second road crossing the Alps – through the Brenner Pass – by Septimius Severus, taking over the task of Via Claudia Augusta; the new road was finished twenty years later, in 215, by Caracalla. However, before that time the Brenner Pass route was already in use as a mountain path;[154] the milestones indicate the intensive maintenance of the road in the third and fourth centuries.[155]

Mountain roads were usually narrower than standard roads – mostly no more than 3 m wide, often only 1.5 to 2 m and only paved over a width of 1 to 1.5 m[156] – and not always suitable for wagons.[157] Where wagon traffic was possible, stone walls were erected to give some safety. In the Alps and Balkan regions artificial ruts have been found for the guidance of the wagons on the road (Fig. 1.4).[158]

It has to be said that the Romans tried as hard as they could to make voyages through high mountains relatively comfortable; they planned their roads as straight as possible and kept the inclination as gentle as possible.

Figure 1.4 Guidance ruts in Illyria and the Alps. Bulle Taf. 3 and 25. München: Verlag der Bayerischen Akademie der Wissenschaften.

Figure 1.5 Pole-brake rut on Blackstone Edge, North Yorkshire. Davies 82. Stroud: Tempus Publishing.

Sometimes it demanded much investment: west of Aosta, no less than eight constructions have been built along a short distance.[159] In less populated regions, there was less construction, or none at all. In the north of England (Blackstone Edge, North Yorkshire) there is a road with a deep rut in the centre, caused by pole-brakes used by carters (on a gradient of 20 per cent) (Fig. 1.5).[160] Besides that, zigzag curves were also used when the slopes were too steep. A zigzag construction is found in Lincoln, where pedestrians could use a stairway, but wheeled traffic had to approach the south gate of the upper city along a wide road diversion.[161]

Nevertheless, travelling through high mountains was no pleasure. Strabo gives an extended description of discomfort and dangers for travellers in his *Geographia*: dizzying heights and the danger of falling into a canyon,[162] in combination with the danger of banditry. Not only for civilians, also for hardened soldiers, a march through the Alps was a difficult task.[163]

The construction of marsh roads

No less complicated was the construction of roads through swamps. During the construction of Via Appia the Romans found the Pontine Marshes to be an obstacle along their way.

In marshy soil, wood was the most important construction material. In prehistoric times log roads were already being constructed, amongst others in the Dutch province of Drenthe.[164] The Romans called log roads *pontes longi*, 'long bridges'. Lengthwise wooden logs were laid down; widthwise wooden

Figure 1.6 Valkenburg a/d Rijn, road on coffer-dam. Colenbrander 2003, 41. Rotterdam: MUST.

beams, logs or boards were placed upon them. These *pontes longi* were for military purposes; this confirms that new roads in newly conquered regions in the first instance were realised by the army.[165] Tacitus describes the role of *pontes longi* in Germania.[166] The *limes*-road along the north frontier also contained this type of road, but through the course of history more solid roads with a gravel surface replaced the log roads.[167] Roman log roads were, in fact, temporary roads.

Another way of crossing a swamp was to construct a dam using any available material. Such a dam was called an *agger* and was also constructed for military purposes, e.g. the catastrophic campaign of Varus in Germania.[168] When a dam was placed on pillars, the Romans spoke about a *pons*.[169] Catullus speaks about a *pons longus* over marshy ground in the neighbourhood of a *colonia*.[170]

In Valkenburg a/d Rijn (west of Leiden, Netherlands) a reconstruction of the Roman road to Voorburg has been built and shows a road mass of sand and clay, resting on wooden pillars (coffer-dam), with a *pavimentum* of gravel (Fig. 1.6). In Utrecht/Veldhuizen, such a construction has also been found.[171] Here we have a solid, permanent road like an *agger*, built of locally available material. Such roads are also to be found in the swampy ground of Trastevere (Rome) and Ostia.[172]

Along the road: drainage canals, sidewalks, stepping stones and milestones

The physical edge of a road is not the edge of the road surface, but the drainage canals beside the road (*sulci*). Everywhere along the roads, these canals are found; the width can be some 10 feet (approx. 3 m); the depth can be equal to the depth of the road mass or deeper, 10 feet.

The position of the canals in respect of the road varies. Forbes gives many examples of roads where the *sulci* are situated directly along the road, e.g. part of the road found in Houthem (east of Maastricht, Netherlands), which was part of the Bavay–Cologne highway.[173] On the same road, in the neighbourhood of Liberchies (Belgium), the situation occurs where the distance between the edge of the road surface and the drainage canal amounts to units of 10 m.[174]

In some places, amongst others Via Flaminia, the road is bound by upright edges (*umbones*) with a height of 30 to 50 cm. For drainage, holes were created.[175] These upright edges could have a safety function or enable a horse or other riding animal to be mounted. We know that C. Gracchus placed special stones for horsemen to mount or dismount,[176] but no archaeological evidence of such stones has been found.[177] In some places wider sidewalks, *crepidines*, were situated along the road.[178]

Also ascribed to C. Gracchus is the systematic introduction of milestones,[179] but the oldest milestone found dates from 189 BC, along Via Appia.[180] Nowadays some 5000 milestones or fragments of milestones have been found,[181] giving abundant information concerning the construction and maintenance of the roads. They give not only the distances to the closest cities, but also the persons responsible for the repair and maintenance of the roads. Augustus erected the *Miliarium Aureum*, the 'Golden Milestone' at the Forum Romanum; on its base, the distances to the city gates were written.[182] Outside Rome, the distances to the nearest cities or villages were counted from the *pomerium* of the Servian Wall.[183] One can say this 'Golden Milestone' was the navel of the Roman road-system. In founding Constantinople, Constantine erected as pendant of the 'Golden Milestone' the *Milion*, a dome set on four columns.[184] Furthermore, there were other cities with a central milestone, including Ephesus (during the Republic the central point of the road-system of Asia Minor)[185] and Augsburg.[186]

Milestones usually had a maximum height of 3 m and were placed 2 to 3 m from the edge of the road surface, set on a square base.[187] The inscription was, of course, engraved on the road side of the milestone and gave, beside the name of the road builder or restorer, the distance in miles[188] or – from 202 AD – in *leugae*.[189] Milestones in Hellenistic provinces also gave the mile as linear measure, but bilingually: beside the distance, the name of the following city in Latin and in Greek.[190] Herms, since archaic times part of the Greek street scene, sometimes mentioned distances and functioned as milestones.[191]

In some important cities road-signs (*itineraria*) were placed; these gave information about stopping places along the roads. In Autun and Tongeren such *itineraria* have been found; the road-sign at Tongeren was originally an octagonal basalt column next to the west gate. Also in this *itinerarium* the distances are given in *leugae*.[192]

Constructions

In the Roman Empire four constructions for road building were in use: bridges, dams, retaining walls and tunnels in mountains. River-crossing tunnels and viaducts[193] were unknown. Aqueducts, however, were mostly underground or crossed the landscape by arched galleries; in fact viaducts were common for water supply. Here a short description of Roman constructions is given. It is noteworthy that particularly in Italy constructions were realised thanks to the abundance of stone and the dense road-system.

Bridges, dams and retaining walls

In Latin different words were used for bridges and dams. The Latin word for 'bridge' is *pons*; the word for 'dam' is *agger*. The Greek word *γέφυρα*, however, signified both 'bridge' and 'dam' at first, but later it developed the meaning of 'bridge'.[194]

One can only speak about a *pons* when a span of a river or valley is part of a road. When an aqueduct crosses a river or valley, this word is not used. The imposing Pont du Gard, part of an aqueduct, is a *pons* because pedestrians also use the construction.[195]

The Romans were not the first who were concerned with building bridges. The Etruscans already had the knowledge of bridge constructing, Vulci being one example. Besides that, the kingdom of Pergamum also built bridges; according to van Buren the influence of Pergamum was most important for the Romans.[196] The fact that in Italy there were more rivers than in most Hellenistic regions was the reason that the construction of bridges played an important role in the realisation of roads. The introduction of the arch construction and mortar created the possibility for Roman engineers to construct wider spans. So we know that some bridges in the course of history were enlarged or replaced by more durable material (stone). Ponte di Nona (the name is derived from the ninth milestone in Via Praenestina, east of Rome) was originally a bridge with one span (arch) and was later enlarged to a longer bridge with seven spans, with a total length of about 72 m; the width of the road is 20 feet (6.25 m); including the *crepidines*, the total width is 10.20 m.[197] We find an equal width at Ponte Amato, also in Via Praenestina.[198] We can find an example of a bridge replaced by a stone one over the Tamega river, at Chaves (in north Portugal).[199]

River bridges contain three parts: the arch or arches over the centre of the river and the ascent and descent roads. To construct bridge surfaces as horizontal as possible, on many bridges the central arch was the same height as the flanking arches. When it was necessary to construct the central river-crossing arch or arches higher for crossing swollen rivers during some periods, the central part of the bridge was higher than the ascent and descent slopes.[200]

In northern Europe, where there was less traffic, there were simple wooden bridges and even fords. Along the roads in the Netherlands, there must have been an abundance of wooden bridges. Longer bridges, crossing wide rivers, are found in Cuijk and Cologne amongst other places. These bridges had stone pillars and a wooden carriageway.[201] In Britain, some thousands of (little) bridges are found, most of them in northern England, owing to a greater availability of stone.[202]

Dam roads (*aggeres*) in marshes have already been mentioned above, p. 19. Roman engineers also constructed dams in rivers, straits[203] and valleys, keeping the surface as horizontal as possible. Two walls were erected (retaining walls); the space between them was filled with sand, earth or mortar, on top of which the road surface was built. A good example of such a construction is the so-called Ponte del Diavolo in Via Salaria (north-east of Rome) (Fig. 1.7). The height is 9.80 m and the length is 30 m, retained by seven buttresses, the width being 5.1 m.[204]

Besides that, these constructions were also used as approaches to bridges, like the bridges in Via Trajana.[205] Via Nomentana in Rome, immediately outside Porta Nomentana, was also provided with well-preserved retaining walls and buttresses.[206] Roads were, of course, equipped with culverts for a well-functioning water management; amongst others, in the case of Via Flaminia at Pieve Fanonica, such a culvert is very well preserved.[207]

We can also find retaining walls in places where the course of a road follows a slope. A wall was built against the slope, supporting the road surface. This is used in Via Salaria, between Rome and Alto Velino.[208] Figure 1.8 shows the situation in the neighbourhood of Aosta, where retaining walls were also in use.[209]

Figure 1.7 Via Salaria, Ponte del Diavolo. Laurence 1999, 198. London: Routledge.

Figure 1.8 Road at Aosta with ruts and milestone. Grewe 12. Mainz: von Zabern.

Tunnels

The Romans were not the first tunnel-constructors. At the beginning of the first millennium BC tunnels for water supply had already been built in Persia. In Palestine, in Megiddo amongst others, a tunnel system was built between 871 and 852 BC. Another was made in the sixth century BC in Athens, and the Eupalinos tunnel on the island of Samos is from the same century. This tunnel had a length of 7 stadions and a height of 8 feet.[210] A particular feature of this tunnel is that two teams of constructors started from both sides, meeting each other in the centre of the tunnel. One was able to estimate the linear direction of the tunnel.[211] Mistakes also occurred: in the tunnel of the Roman engineer Nonius Datus at Saldae (North Africa), the teams missed each other.[212] These tunnels were for water supply and they are not discussed further in this book.

The first tunnels for traffic were constructed by L. Cocceius Auctus,[213] a contemporary of Agrippa (last decades BC). These tunnels are in the environs of Naples.[214] In Cumae, there are two tunnels: the Cocceius tunnel[215] (with a length of 1 km) and the Cripta Romana. Further, there is the Cripta Neapolitana between Pozzuoli and Naples; this tunnel had a length of 705 m and a width of 3 to 4 m. This tunnel is also mentioned on the Peutinger Map (Fig. 1.9).[216] In the neighbourhood of Naples there was also the Grotta

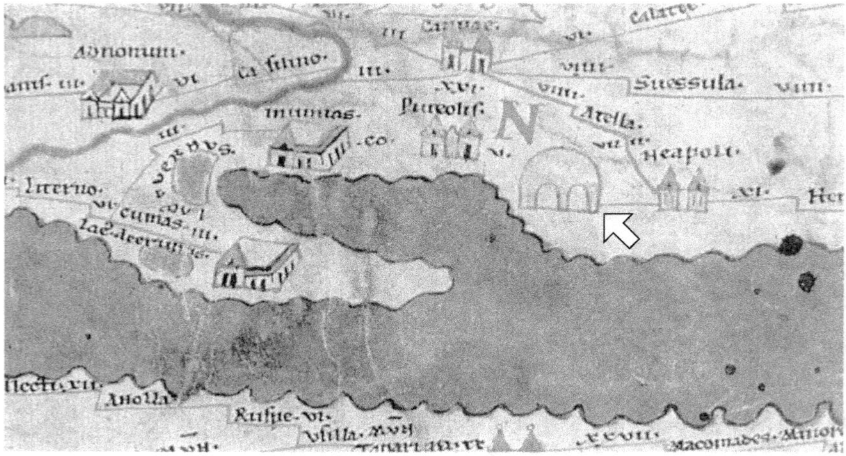

Figure 1.9 Tunnel on the Peutinger Map. Grewe 125. Wien: Österreichische Nationalbibliothek.

di Seiano (date unknown)[217] and the Chiaia di Luna tunnel on the island of Ponza. The Grotta di Seiano had a length of 780 m and a width of 4 to 6 m. Strabo mentions that traffic could pass and overtake each other.[218] The tunnel at Ponza was smaller: 170 m long and 2.2 m wide.[219]

There were also tunnels in the Furlo Pass, also named the Petra Pertusa on Via Flaminia, between Rome and the Adriatic Sea, during the time of Vespasian,[220] with a width of 4.5 to 6 m,[221] and on the road between Vienne and Grenoble with a width of some 2.46 m. The road surface had ruts.[222]

There were about twenty tunnels in Italy[223] but their construction was limited as much as possible; not only were tunnels expensive to construct, but travellers preferred not to use them. Seneca writes to his friend Lucilius about the Cripta Neapolitana – comparing travelling with wrestling athletes, oiled and powdered:

> *A ceromate nos haphe excepit in crypta Neapolitana. Nihil illo carcere longius, nihil illis facibus obscurius, quae nobis praestant non ut per tenebras videamus, sed ut ipsas. Ceterum etiam si locus haberet lucem, pulvis auferret, in aperto quoque res gravis et molesta: quid illic, ubi in se volutatur et, cum sine ullo spiramento sit inclusus, in ipsos, a quibus excitatus est, recidit? Duo incommoda inter se contraria simul pertulimus: eadem via, eodem die et luto et pulvere laboravimus.*

> The anointing with which we began was followed by the sand-sprinkle in the Naples tunnel. No place could be longer than that prison; nothing could be dimmer than those torches, which enabled

> us not to see amid the darkness, but to see the darkness. But even supposing that there was light in the place, the dust, which is an oppressive and disagreeable thing even in the open air, would destroy the light; how much worse the dust is there, where it rolls back upon itself, and, being shut in without ventilation, blows back in the faces of those who set it going! So we endured two inconveniences at the same time, and they were diametrically different: we struggled both with mud and and with dust on the same road and on the same day.
> (Sen. *Ep.* 57.1-2)

The tunnel of Clivus Tiburtinus

The tunnel of Clivus Tiburtinus in Tivoli is a special one, where a branch of Via Tiburtina runs beneath a building complex.[224] Although this is not a mountain but a sanctuary passage, it should, however, be called a tunnel. This tunnel is called *via tecta* and takes in part of the temple complex west of Tivoli, including the temple of Hercules Victor and a theatre. Because the temple complex is erected against a slope, the road is a gradient (*clivus*).[225] The *via tecta* has a length of about 90 m and a width of about 8.5 m (28 feet)[226] and included vaulted market halls. There were shops along it. To give light – very necessary here – there were four skylights in the ceiling of the tunnel (Figs 1.10 and 1.11).[227] On the plan, to the left of the theatre and the temple the *via tecta* is visible, running at an oblique angle. The four rectangular structures are the skylights.

The *via tecta* is not constructed to shorten a distance; traffic problems did not play a role here. Another trace of Via Tiburtina ran south of the complex to Tivoli, to Porta Maggiore.[228] The construction of the *via tecta* has, in this case, a religious aspect. Via Tiburtina was part of a pastoral landscape; not only was this sloping road used by shepherds, but Hercules was also a god of shepherds and markets.[229]

Widths of roads

The potential to sustain traffic flow is directly dependent upon the width of the road. To prevent traffic congestion, a wider road is more efficient than a narrower one.

In the previous section, the width of roads is in some cases already mentioned as being juridical as well as practical. Not all roads had the same width, meaning that not all roads were suitable for all kinds of traffic. The planning of the width depended on two factors: the juridical status of the road and local circumstances. Only under the most extreme conditions was a road built narrower than as the law prescribed.

Figure 1.10 Tivoli, reconstruction of the temple complex, with *via tecta* on the left. Santillo Frizell 32. Roma: Presso Beni Culturali.

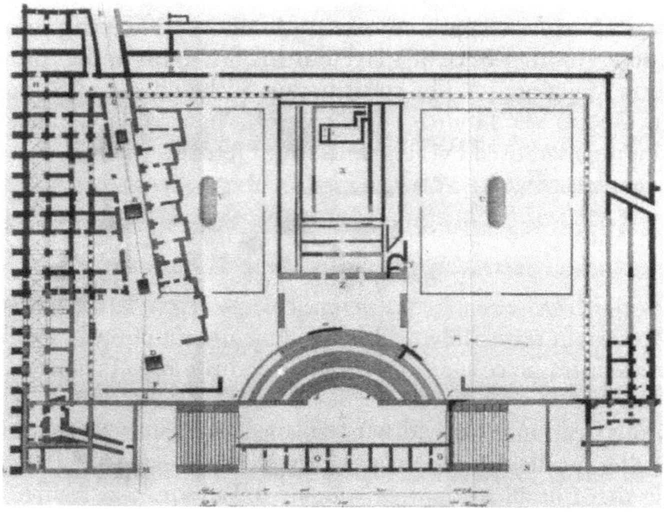

Figure 1.11 Tivoli, plan with skylights. Basso 47. Roma: l'Erma di Bretschneider.

The minimum width of roads: juridical aspects

The Romans used standard dimensions for roads. In the previous section it has already been mentioned that a *via*, a road suitable for all kinds of traffic, had to be a minimum width of 8 feet; were bends a minimum width of 16 feet also prescribed.[230] This width was enough for wagons to pass or overtake each other. It is generally assumed that this standard was already mentioned in the so-called Law of the Twelve Tables (*c.* 450 BC), following Gaius' statement.[231] This standard was valid for all *viae*: not only for *viae publicae*, but also for *viae privatae*, according to Festus.[232] In the same sentence it is mentioned that there were no hard and fast dimensions for *viae publicae*;[233] probably there was no maximum width. Under certain circumstances, however, a *via* could be narrower than 8 feet;[234] maybe here *viae privatae* are meant, having less traffic than *viae publicae*.[235]

The 8 feet norm was not in force for the *actus* and the *iter*. An *actus* had a width of 4 feet and an *iter* of 2 feet.[236] According to Javolenus, there was no prescribed width for these roads. Because they were used less, there were fewer problems in relation to their width. In the case of a difficulty with a *via*, the 8 feet norm was legislative.[237]

The width of roads in geodesy

The 8 feet norm also occurs in geodesy. It is the minimum width of the so-called *subruncivi*, the narrowest roads between the plots of allocated land.

The hierarchy of roads in allocated land (with a chessboard pattern) is fixed by their width. At the top, are the *cardo* and *decumanus* with a width of 40, 30, 20, 15 or 12 feet; thereafter the *actuarii* (12 feet) and, lastly, the *subruncivi* (8 feet). These numbers are laid down by Hyginus Gromaticus (the Surveyor).[238] They are derived from the surveyor's standard unit of 120 feet and multiples of 3 and 4; 120 is divisible by all these numbers.[239] The *cardines* and *decumani* would have been *viae* (*publicae*), accessible for everyone and wide enough for wagons to pass and to overtake. The *limites*, being lower in the hierarchy, were narrower, with a minimum width of 8 feet as mentioned above. In the *Liber Coloniarum* there is a passage which describes the hierarchy of land allocation.[240]

Pliny the Elder gives a description of land allocation: he advises on planning a *cardo* and a *decumanus* in a vineyard with a width of 18 feet, *limites* with a width of 9 feet and a *quintana*, a 'fifth path' by every fifth vine.[241] These measurements are wider and conflict with Hyginus' dimensions, but they are still multiples of three.

Otherwise, not the surveyors but the principals (the *imperator* or local authorities) fixed land allocation dimensions and road widths.[242]

The actual width of roads

Did these standard widths also actually apply?

By means of excavations and research on existing Roman roads or their tracks, we can conclude that *viae publicae* had wider dimensions than the prescribed 8 feet norm of Hyginus Gromaticus, Varro and the jurists. In classical literature, there is no mention of maximum widths, but in the case of geodesy there are prescribed widths of roads and paths within the hierarchy of land allocation. With the lack of prescribed maximum widths, roads had a variety of widths.

According to the scholars H.-C. Schneider, Bender and Pekáry, *viae publicae* had a minimum width of 5 to 7 m (15 to 20 feet), each side flanked by a strip of 3 m, which was held to be free.[243] Pekáry states that a *via publica* had a standard width of 40 feet; half of it was taken up by the paved road surface and the other half by the unpaved strips. After these strips came the sepulchral monuments. These numbers are obviously not valid for *viae privatae, viae vicinales* and *viae publicae* in mountains.[244] Pekáry also gives the measurements of roads in Italy (4–7 m), Gaul (5–6 m) and Pannonia (5.5–10.5 m). Via Egnatia in Greece has a width of 5.7 m.[245]

The famous Via Appia, commonly known as a long, straight, wide and well-paved road, sometimes had a minimum width of 8 feet; elsewhere 10, 14 to 18 feet and in front of city gates 30 feet.[246] At Emona (Pannonia) a *via publica* had a width of 50 feet;[247] in some mountainous regions no more than 6 feet with the necessary crossings.[248] Because there was hardly any traffic in these mountainous areas, and certainly no traffic congestion, one has to assume that this caused few logistical problems.

In Hyginus' work numbers occur which give – when seen as road widths – enormous widths such as 60 feet (18 m) at Nuceria, 80 feet (24 m) at Naples, 100 feet (30 m) at Capua and even 120 feet (40 m) at Nola. According to Pekáry, it is not the road widths themselves which are meant here, but plots of land where the *limites* coincide with the road; when in the case of land allocation a main road already existed, the road could coincide with a *cardo* or *decumanus*.[249] Forbes also mentions a figure of 80 feet,[250] maybe based on these numbers.

Nevertheless, roads *are* found with a width of tens of metres, i.e. between the boundary ditches of the road. Excavations show that the Bavay–Cologne highway had a road surface with standard dimensions (5–7 m), but which was flanked by unpaved strips with a width of tens of metres, marked by ditches. In Germania Inferior, road sections were found with a width of 40 to 50 m.[251] At Liberchies (Belgium) the axes of the ditches ran 20 and 17 m from the road's axis; we see the same dimensions in the Amiens–Senlis road and the north exit route of London. In these cases, horsemen, pedestrians and herds used the unpaved strips.[252] In Britain, there were paved sections of parallel roads, amongst others between London and Colchester; only the central road

Table 1.1 Road widths of the Bavay–Cologne highway

Location	*Paved section (gravel)*	*Unpaved section*	*Total*
Liberchies	6 m	13–16 m	about 37 m
Maastricht	9 m, later 29.20 m	22.20 m (south side)	29.20 m
Rimburg	7–7.75 m, later 9–12 m, finally 10.75–11.20 m	26 m	finally 37 m
Gross Königsdorf/ Quadrath	5–7 m	17.80–19.80 m	24.80 m

shows traces of intensive traffic.[253] Traffic congestion on these roads seems to have been improbable; there was space enough to pass. Table 1.1 shows the different widths of the Bavay–Cologne highway at various points.[254]

Paved sections of a road with a width of tens of metres are, of course, found as forecourts before city gates. Examples include: the width in front of Burginatiumtor in Xanten at 12 m. The road at the gate in Nîmes, Porta Augusta, has a width of some 9.50 m.,[255] a gate with four passages. The width of the road coincides with the widths of the central passages of the gate, including the central pillar; the whole forecourt had a width of 20.30 m including the *crepidines* for pedestrians.[256] Other cities like Autun and Aosta have gates with forecourts about 20 m wide (60 to 70 feet).[257] A good example of a gate with a wide forecourt can be found in Timgad (Algeria) (Fig. 1.12).

Widths of streets

The situation regarding streets in cities is the same as that for roads: there were streets wide enough for parking wagons, without affecting passing wagons, and there were streets too narrow for a wagon at all.

In general, one can conclude that narrow streets occur in older cities, chaotically planned cities and cities with a hot climate. Wide streets occur mostly in later cities, cities situated in a cooler climate and cities planned with a chessboard pattern. Some cities in the south obviously also have wide streets and some streets in northern cities are narrow.

In Greek areas, there is a marked difference between main roads and other roads. The main roads, λεωφόροι, have a width of 20 ells (about 10 m) or more and the other ὁδοί 8 ells (about 4 m).[258] In cities, we also see the same pattern. Priene, a mountain city in Asia Minor, is built according to the Hippodamic plan: a plan with a chessboard pattern. The main streets have a width of some 7 m, the side streets 3.20–4.50 m. Such widths also occur in Magnesia (4.50–5 m) and Selinus (3.60–5.40 m). In Alexandria,

Figure 1.12 Aerial view of Timgad, at the upper side the gate with forecourt. Scullard 1964, ill. 194. Amsterdam/Brussels: Agon Elsevier.

normal streets had a width of 6–7 m, as wide as the main streets of Priene. In some Hellenistic cities, there were even wider main streets. In Palmyra, the main street had a width of 11.50 m, flanked by colonnades with a width of 5.50 m.[259] In Alexandria, according to Strabo, the widest streets – or rather avenues – had a width of more than 30 m. Here were also colonnades.[260] The widest streets were those at Miletus, with a width of more than 40 m. They were constructed not for traffic purposes, but for processions. Miletus had a new Hippodamic street grid in 479 BC, where the wide avenues functioned as a co-ordinated system. Apart from this design, unadapted further because they occupied too much space,[261] a development is visible – according to Hoepfner and Schwandner – of a city with larger housing blocks and narrow streets altered to a city with smaller housing blocks and wider streets. The reason for this development was to obtain a better traffic flow.[262] This was also the case with the wide streets in Rhodes and Alexandria connecting the harbour basins, but the extremely wide streets of 30 to 40 m seem to me to be designed as a status symbol.

An example of such a southern city with wide and narrow streets is Pompeii. The oldest part of the city (districts VII and VIII) shows a disorderly pattern of small, winding streets; the other districts are planned and have straight streets. In these districts we also find the widest streets: Via di Mercurio (the former *cardo*) has a width of 32 feet; Via di Nola and Via dell'Abbondanza (the *decumani*) 28 feet; and Via di Stabia (the later *cardo*) about 24 feet. The narrowest (side) streets have a width of 8 feet (2.40 m).[263] Such streets are too narrow for two-way traffic[264] and one-way traffic had to follow certain routes.[265]

In Rome, there were hardly any wide streets. According to Stambaugh there were only four streets with Varro's status of *via*: Via Sacra, Via Nova,[266] Via Lata and Via Tecta.[267] In which case, all other streets of Rome would have been narrower than 8 feet (2.40 m). According to Homo, Via Nova and Via Lata were wide enough to attain the status of *platea*;[268] F. Kolb even gives Via Nova a width of 30 m![269] Beside these avenues, there were hardly any streets with the name of *via*.[270] They were not *viae publicae*; they ran *between* cities and not *within* cities where the city council was responsible for the streets.[271] The majority of the streets were narrower than was officially permitted.[272] Attempts to create wider streets failed, after the catastrophic fires under Nero and Commodus.[273]

In contrast, let us take a look at the wide streets in northern Europe. Colchester had streets with a width of 17–20 feet (5–6 m), including one of 31 feet (9 m).[274] In Bavay a variety of widths are found, between 3.20 and 6.80 m, excluding the sidewalks and ditches; including these the streets have a width of 10–12 m.[275] Still wider are the streets of Xanten; only the driving-roads had a width of 10–12 m. The sidewalks were flanked by colonnades with a width of about 4 m. There were no ditches along the street, but underground sewers to the city gates discharged rainwater.[276] One can conclude that the development from narrow to wide streets in Hellenistic cities is continued in Roman *coloniae*, with Xanten as the final result after centuries of planning development.

Summary and conclusion

The Roman Empire had an excellent road-system, its construction and width making it suitable for all kinds of traffic, certainly compared with the preceding prehistory and the following Middle Ages. Through traffic could make use of straight running roads with a minimum width of 8 feet nearly everywhere and usually more. Only in a few places – in mountainous regions – was the width less than 8 feet, but a width of 18–24 feet (5–7 m), enough for two-way traffic, was standard. Beside the paved surface of a road (*pavimentum*) there were drainage ditches (*sulci*) and sometimes unpaved strips between the road surface and the *sulci*. To construct a road as straight as possible, the Romans built artificial constructions like dams, bridges and tunnels.

In passing through cities, through traffic used the main streets; these were usually no narrower than the roads outside the cities. New cities planned by the Romans (*coloniae*) had a chessboard pattern, based on the *castra* model, including wide streets for two-way traffic, like Xanten. Older cities, like Pompeii, had a pattern of narrower streets, suitable for one-way traffic.

Principals and road authorities

Introduction

A description of the road-system is impossible without paying due attention to those who controlled road building and maintenance. Civilians and authorities alike realised and financed the construction of the roads. There were several bodies responsible for maintenance. A badly paved road is a problem for a steady stream of traffic, so roads needed regular maintenance. However, we know of some cases of badly paved roads and collapsed bridges, resulting from civil wars or neglect.

In this section, a short description will be given of the diverse magistrates, responsible for the planning, building and maintenance of roads and streets.

Politicians as principals

It remains a point of discussion that censors – beginning with Appius Claudius Caecus, the man who gave his name to Via Appia – were actually concerned with road building. Mommsen confirms this.[277] Livy mentions the censor C. Flaminius, who built Via Flaminia in 220 BC.[278] Hinrichs states that censors were concerned with road building, because of the expense, until the Second Punic War; after which, roads were financed by booty from the wars with Carthage and Spain.[279] Radke and Pekáry are of the opinion that censors had no authority to make decisions on road building.[280] It is a fact that no milestones are known to allude to *viae censoriae*; in the law-books, this classification is unknown.[281] According to Ulpianus a *via publica* could be constructed only on state land and a censor did not have permission to expropriate private possession.[282] For that reason, in 174 BC censors were obliged to build streets in Rome; this was city land, not *ager publicus*.[283] It is significant that Livy mentions censors in the case of the first paved roads and streets; later in history, censors are no longer mentioned by Livy as road builders.[284]

The politicians concerned with road building in Italy were for the most part consuls, who also gave their names to the roads. So Via Aurelia (Rome–Etruria) got the name of C. Aurelius Cotta, consul in 241 BC, and

Via Aemilia (Rimini–Piacenza) of M. Aemilius Lepidus, consul in 187 BC.[285] Occasionally, a road was built by a praetor;[286] according to Radke the stretch Formiae–Capua on Via Appia was built by Appius during his praetorship in 295 BC.[287] Via Domitia (southern Gaul) was built by proconsul Cn. Domitius Ahenobarbus in 121 BC.[288]

The most important information concerning the functions of the principals is given by milestones. In 252 BC[289] we already find along a road in Sicily an inscription of Aurelius Cotta; an undated inscription mentions Appius as consul.[290] Some other examples of consular roads (*viae consulares*) include Via Aemilia, 187 BC[291] and 175 BC,[292] Via Postumia, 148 BC,[293] Via Latina, 127 BC[294] and Via Caecilia, 117 BC.[295] When Rome became an Empire, the emperors became the principals. Besides Via Domitiana[296] we know of Via Trajana, a branch of Via Appia, also finding its terminus in Brindisi.[297]

Outside Italy, road names – whether with names of politicians or not – scarcely occur and those that do, dating from the Republican era, are not mentioned earlier than the first century BC; thus Via Salaria is named as such for the first time by Cicero,[298] who also gives the first references to the names of Via Domitia,[299] Via Appia[300] and other roads.[301]

Viae militares

The main reason for the construction of a road was to facilitate troop movements and rapid communications.[302] For that reason it is generally assumed that the army was also a principal.[303] There is mention of *viae militares*, but a clear definition is lacking. According to Ulpianus, they were part of *viae vicinales*; they occurred by rivers and cities. He says nothing further about the specific military character of them.[304] Pekáry and H.-C. Schneider agree with Ulpianus' statement; Schneider also gives other views of scholars, who see *viae militares* as roads in the frontier provinces, important for the army.[305] Chapot, however, states that *viae militares* were provisional, short roads, constructed for strategic purposes.[306] In any case, *viae militares* occurred in Dalmatia and north Greece; Via Egnatia, the most important east–west connection, is described as a *via militaris*.[307] This is, however, not a '*chemin en général assez court et peu soigné*', as described by Chapot, but an important highway. In this context, *viae militares* were important roads with accommodation for travellers. Suetonius mentions *viae militares* used by Augustus to provide postal traffic between provinces[308] and André and Baslez speak of *viae militares* at Autun.[309] I agree with Pekáry that *viae militares* were not special-category roads, but roads with such an important strategic function that the army supervised them. But there is no complete definition of *viae militares*.[310]

Anyway, in newly conquered regions the army built new roads, provisional or permanent, as soon as possible. The *limes*-road along the Rhine was already built by 20 or 19 BC,[311] in a newly conquered area; in Gaul, roads must have been built earlier, upon Caesar's conquest. The *pontes longi* in Germania, described in the preceding section, can be also seen as *viae militares*, because they were built by and on behalf of the army.

Roads as an instrument for colonisation and reclaiming land

Not all roads were built to stimulate rapid troop movements and communications. The construction of roads to unlock allocated land had a merely civil purpose. On p. 6 I mention that C. Gracchus stimulated road construction in his *Lex Sempronia viaria*. It is probable that these were roads for allocation and roads running to the new *coloniae*.[312] The following inscription of praetor P. Popilius Laenas from 132 BC shows the recruitment of workers, who could acquire a plot after the completion of Via Popilia in Lucania; Via Popilia was a road in a peaceful Roman region, so in this case there was no military purpose.[313] P. Popilius Laenas was an opponent of C. Gracchus; he probably attempted by means of his own colonisation projects – similar to those of Gracchus – to diminish the growing political power of Gracchus.

Curatores viarum

The organisation of road maintenance was fragmented. Besides censors, praetors and consuls, aediles were also known as road builders, but during the second century BC their capability was restricted to the maintenance of roads and streets up to one mile outside the boundary of the city (the so-called *pomerium*); Livy tells us that in the case of Rome, aediles had capability up to 10 miles outside the boundary.[314] In Asia Minor, possibly a *quaestor* repaired a road.[315]

To solve the problems of this fragmentation, in the first century BC some attempts were made to create more structure in road maintenance. In an inscription dating from 72 or 70 BC we see a *curator viarum e lege Visellia*[316] and in 50 BC the tribune of the people C. Scribonius Curio introduced the *lex viaria,* a road bill, but it probably failed.[317] The maintenance of roads also failed; Cicero made complaints about the bad condition of Via Appia.[318] Via Flaminia was supervised in 65 BC by a *curator viarum*, C. Minucius Thermus.[319] *Curatores* also engaged in the construction of bridges; the Pons Fabricius was built by the *curator viarum* L. Fabricius in 62 BC.[320]

After the assassination of Caesar, more attention was paid to road repair. Senators repaired roads at their own expense;[321] the senators C. Calvisius Sabinus and Valerius Messala gave the order to repair Via Latina, Valerius Messala Via Appia.[322]

During the reign of Augustus, from 27 BC onwards, a campaign was initiated to tackle effectively the bad condition of the road-system.[323] He embarked upon extended projects with Via Flaminia (with a triumphal arch in Rimini) and bridges in Rome; barring the Pons Milvius and the – as yet unknown – Pons Minucius, their condition being too bad.[324] Further he disposed of large amounts of money to triumphers for repairing roads.[325]

To simplify the decision-making, Augustus renewed the function of *curatores viarum* in 20 BC. The new *curatores viarum* came from the ranks of senators and knights and acquired the assignment of two lictors. Their task included the completion of new roads and their maintenance; further, they had to supervise the condition of the existing roads.[326] Because on some milestones dating from the reign of Augustus the abbreviation *ex s.c.* (*ex senatus consulto*) is not mentioned, Eck concludes that the installation of the *curatores viarum* made repairs of roads possible.[327]

The *curatores viarum* were congregated in a *collegium*, for example the *collegium* founded in 11 BC for the *curatores aquarum*.[328] Eck states that they initially supervised all roads, certainly in the period of Augustus; he names two *curatores viarum*, P. Paquius Scaeva and C. Propertius Postumus. Later they supervised part of the road-system and in most cases only one road. There is discussion on the number of the *curatores viarum*; according to Eck fewer than eight, but more than two under Augustus; after Augustus there were eight *curatores viarum*.[329]

Eck gives a general overview of road repairs in Italy in the period from Augustus to Alexander Severus.[330] How necessary these repairs were is shown by the following inscription (Fig. 1.13) from the year 123/4: Via Appia was no longer accessible to traffic.[331]

IMP · CAESAR
DIVI · TRAIANI
PARTHICI · F · DIVI
NERVAE · NEPOS
TRAIANVS · HADRIANVS
AVG · PONT · MAX · TRIB
POT · VII · COS · III p. C. 123
VIAM · APPIAM · PER
MILLIA · PASSVS
XVƉCCL · LONGA
VETVSTATE AMIS
SAM · ADIECTIS
HS · XI · XLVII AD
HS · ƉLXIX C QVAE
POSSESSORES · AGRO
RVM · CONTVLERVNT
· FECIT ·

Figure 1.13 *CIL* IX 6075. Berolini: Reimer/de Gruyter.

> Emperor Hadrian . . . has repaired Via Appia for the distance of 15.75 miles, impassable by old age, by adding the sum of 1,470,000 sesterces to the sum of 569,100 sesterces collected by the land owners.

It is remarkable that the *curatores viarum* only had the supervision of roads in Italy. North of the Po region there is no evidence of the existence of *curatores viarum*. It is generally assumed that there were no tasks for them outside Italy.[332] In the provinces, the proconsuls, the praesides and legates were responsible for the maintenance of roads.[333]

The problems regarding road repairs and maintenance were not solved by the installation of the *curatores viarum*. In 21 AD Cn. Domitius Corbulo condemned the malfunctioning of the *curatores viarum* and accused them of fraudulent practices. In 39 there were also complaints.[334] During the reign of Hadrian, the bad condition of Via Appia was well documented. Maybe this was a consequence of the completion of Via Trajana, which took over a major part of the traffic, so there was less attention paid to Via Appia. Other roads were also in bad condition when repairs became an urgent matter[335] and archaeological sources show that some parts of the road surface were in extremely bad condition (Fig. 1.14).[336]

The actual operations were executed by building contractors (*mancipes* and *redemptores*), who, besides the roads themselves, also erected the *stationes*

Figure 1.14 The bad condition of the surface of Via Trajana at Monopoli, near Brindisi. Laurence 1999, 63. London: Routledge.

(*mansiones* and *mutationes*) for the transport organisation (*cursus publicus*). These *stationes* will be reviewed below, pp. 58 and 62. The road workers themselves were contract workers, captives or slaves; sometimes the army was brought into action.[337]

Finally, something about the costs of road repair. In the aforementioned inscription (*CIL* IX 6075) for the repair of Via Appia in 123/4 a total sum is mentioned of 2,039,100 sesterces (1,470,000 from the emperor and 569,100 from the land owners) for a distance of 15.75 miles.[338] The inscription shows that landowners also collected money; it was not only the emperor or the local authorities who financed road repair, but also the people who lived along the road.[339] Of course, such sums could only be collected in densely crowded regions.

Street maintenance in cities

Organising the maintenance of streets in cities was less complicated. In Athens in the fourth century BC there were five functionaries, the so-called ὁδοποιοί, who had slaves at their disposal.[340] Their task was the supervision of the maintenance of road surfaces. They were supported by a body of traffic police, the ἐπιμεληταί, which finally incorporated about a thousand men. Another body was responsible for keeping the streets clean: the ἀστυνόμοι.[341] In 321 BC, after Alexander's death at the beginning of the Hellenistic period, the ἀστυνόμοι were replaced by the ἀγορανόμοι, a common activity in Hellenistic cities.[342]

In Roman cities, the two aediles or *IIviri* were responsible for controlling the maintenance of road surfaces and ensuring that they were kept clean and free of obstacles. In the so-called *Lex Julia Municipalis*, the aediles of Rome are mentioned, supported by *IVviri viis in urbe purgandis* (in the city) and *IIviri viis extra urbem purgandis* (outside the city).[343] It has already been said that the responsibility of the aediles extended to the boundary of the city, including the surrounding land up to one mile, but in the case of Rome up to 10 miles.[344] The aediles were, like their Greek counterparts, also responsible for the conservation of artificial constructions like drain ditches and drinking water supply. In some cities aediles could build new streets, including drain ditches.[345]

In principle, house-owners or tenants had the responsibility of the cleaning and maintenance of the streets before their doors. When the streets were still unpaved, it was not difficult to fulfil this obligation. In the Late Republic and Empire, when paved streets were common, we find two situations: in some cities, the local government took care of the maintenance of the streets, keeping them clean and free of obstacles;[346] in other cities, including Rome, the tenants were obliged to do this. In Rome, the tenants were entirely responsible for the condition of the streets.[347] An explanation of this situation could be that Rome had many narrow, unpaved lanes.[348]

The maintenance of the streets was part of the so-called *munera* or responsibilities. Some juridical texts explain clearly which tasks people had to do, e.g. to maintain the road surface and to keep the drain ditches free.[349] The *munera* were responsibilities of the house- and landowners (*possessores*) and not of tenants, at least not in the time of Papinianus (beginning of the third century AD) and probably earlier.[350] Papinianus also states that it was forbidden to damage streets and drop litter,[351] but the question remains whether the streets were really always clean and unimpaired and we have to assume the negative. Juvenal and Seneca complain about muddy streets.[352] Probably the situation was no different from our roads nowadays.

Streets had not only to be clean but also passable, and it was an obligation (*munus*) to keep them free of obstacles. The aediles ensured that shop-owners did not place their merchandise in the street so that it was blocked. When they did this, the aedilis had the power to wreck the merchandise.[353] Fullers and carpenters had, according to Papinianus, the right to exhibit their merchandise, on condition that it did not block the traffic.[354] In 92, Domitian commanded that shop-owners no longer had permission to exhibit their merchandise on the sidewalks because they blocked (pedestrian) traffic.[355]

Of course, it was forbidden to build structures or balconies which obstructed or blocked the street, even the use of doors opening on to the street.[356] Balconies on the front of buildings were a common sight; they were supported by columns standing on the edge of the sidewalks, so that they did not obstruct pedestrians. In cities like Xanten, whole sidewalks were colonnaded.

Also forbidden was to occupy part of the street for erecting a private house or barn. In the later Roman Empire, there are two laws concerning this kind of 'street theft': it was stated by the emperors Theodosius and Valentinian II that those who annexed a part of a lane or street or a colonnade were fined 50 pounds in gold and the obstruction had to be demolished.[357] In April 397, this law was augmented by the emperors Honorius and Arcadius: anybody who erected illegal buildings at the Campus Martius in Rome was exiled and the buildings were demolished.[358] It is noteworthy that the population of Rome was decreasing at that time; there were fewer inhabitants than in the second century and there was space enough to live. Of course, there were no penalties where there was permission to erect a construction, if it was given by the emperor or the city government.[359]

Summary and conclusion

According to Livy, the first authorities to construct roads were censors, in and outside Rome. Later the main authorities were consuls. During the Late

Republic, province governors gave orders to extend the road-system, and during the Empire this was also the emperor's task. The actual building of roads was done by road builders (*mancipes* and *redemptores*) along with their workers. These were contract workers, captives and slaves; the fact that captives were used shows that the work was very arduous.

There were three authorities concerned with road and street maintenance. At a local level, the aediles supervised the maintenance and cleaning of streets and the freedom from obstacles. Outside the cities, in Italy, the *curatores viarum* controlled the condition of the roads; they had one or more roads under their jurisdiction. Outside Italy, the province governors – proconsuls and legates – oversaw the condition of the roads.

The cost of road maintenance in cities and densely crowded regions was wholly or partly paid by the house- and landowners along the roads. In Rome the inhabitants themselves had to repair and keep the (unpaved) roads and streets clean; in other cities, the city government took care of it. In the case of *viae privatae*, the owner had the responsibility of caring for the condition of the roads.

Laws and authorising bodies could not prevent muddy and badly paved roads; literary, epigraphic and archaeological sources show that sometimes the situation was bad – like today. Travellers could be confronted with muddy and sometimes obstructed and blocked roads and streets; in cities, many traffic restrictions were in force.

Conclusion of chapter 1

The Roman road-system can be seen, without reservation, as the central nervous system of the Roman Empire. Without this road-system the Empire could not be governed and trade and rapid troop movements would have been impossible. Many authors have praised the road-system as helping to create civilisation. The fact that other authors complain about dirty and muddy roads and streets does not diminish their glory.

In the first instance, paved roads were built for troop movements and couriers. It is generally presumed that the main purpose of the famous Via Appia, built from 312 BC onwards and the first paved and planned Roman road, was to create a fast and dependable connection between Rome and Capua, a city which came under Roman influence from the fourth century BC onwards. In three respects, Via Appia was a turning point in the history of road building: it was the first paved road over a long distance (the Etruscans also used paved roads, but they were shorter); it was the first road which took the name of its planner; and, lastly, it was the first straight road, running *linea recta* through the landscape. Later roads followed the same straight concept.

The expansion of the Empire required an expansion of the road-system. From the middle of the second century BC the first roads were built outside Italy: Via Egnatia in northern Greece and Via Domitia from southern Gaul to Spain. At the beginning of the Christian era, the road-system was extended from the Netherlands to Palestine and in the middle of the second century AD from Scotland to the Euphrates river. The road-system had a length of paved highways of some 100,000 km.

The Romans spent a lot of time on the road-system. In the fifth century BC there was already a norm that the minimum width had to be 8 feet (about 2.40 m). The roads were usually twice as wide: 16–20 feet (about 5–6 m) was normal. The roads were well-paved with a solid substructure, with drainage ditches at both sides. Only in less accessible mountain regions were the roads narrower. To get the optimal use, they were built as straight as possible in the landscape with as few as possible differences in altitude. To realise this, many artificial constructions had to be built: bridges, dams and tunnels. For travellers there were milestones and resting places along the road.

Maintaining roads and streets required a lot of money and energy, and badly paved road sections were not repaired immediately. Even a part of Via Appia underwent a period of poor condition during the reign of Hadrian. We must assume that, for the smooth functioning of the Roman Empire, the majority of the roads had to be in good condition.

In the cities the streets were also in good condition; the through roads were no narrower than the roads outside the city. The side streets, however, were narrower. This was the case in the older cities of the Empire, e.g. Pompeii. Newly founded Roman cities, particularly the *coloniae*, sometimes had a street system with streets wide enough for all kinds of traffic on both sides, e.g. Xanten.

In contrast to these *coloniae* and the Hellenistic metropoles, Rome had hardly any wide streets that made two-way traffic possible. The majority of streets were, generally, winding and too narrow. Modern Rome is built upon them, so there is not much (archaeological) information about it. Complaints about streets (amongst others by Juvenal) mainly concern the streets of Rome. Possibly not all the streets of Rome were paved; the clause in the *Lex Julia Municipalis* that inhabitants themselves had to take care of the pavements could be evidence that there were many unpaved streets and lanes.

To conclude, one can state that, in any case, the interurban Roman road-system had the capacity to sustain traffic. The roads were wide enough, the thoroughfares as much as the roads on allocated land. The main streets in the cities were also wide enough. Side streets, however, were narrower and not suitable for wheeled traffic on two sides.

Traffic congestion is caused by two factors: capacity of infrastructure (the width of the road) and the volume of the traffic flow. In the next chapter, road-users will be discussed.

2

ROAD-USERS

Traffic congestion is caused by two factors: the width of the roads and streets and the volume of traffic flow. In chapter 1, we made acquaintance with the infrastructure and its capacity; in this chapter we will learn about the capacity of traffic flow.

As nowadays, in the Roman Empire there were two main categories of traffic: passenger traffic and goods traffic (including transhumance). Besides these categories there were special road-users: the army, the state distribution system administering the state transport (in imperial times known as the *cursus publicus*), travelling emperors with their court (especially in Late Antiquity), transport of animals for amphitheatre games and heavy transport. All these categories will be reviewed in separate sections and attempts will be made to estimate the volume of traffic flow. One problem is that we do not have details or exact dates of the amount of traffic flow. We know the number of soldiers in a legion, so we know, approximately, how many soldiers were on the road when an army of three legions was on its way. But what we know of the anonymous groups of merchants, minor officials and messengers, travelling daily, is extremely scarce. What we know of street crowds we can only conclude from random passages in literature and archaeological sources.

Not all road-users were traffic participants. A major part of daily life took place in the open air, particularly in the Mediterranean regions of the Roman Empire. The streets were full of merchants, beggars, shoppers, children playing, etc.

Passenger traffic

Introduction

There are three categories of passenger traffic: commuter traffic, service traffic and recreational traffic. Nowadays, particularly commuter traffic causes congestion in the morning and evening rush hours. An increasing traffic flow is visible in recreational activities; on Sunday afternoon traffic also becomes busier.

These types of traffic scarcely existed in the Roman Empire, although there tended to be rush hours. Suburbs like ours did not exist and the common (free) citizen, rarely in possession of wagons and riding or draught animals, was unable to make a holiday voyage far outside their domicile. This situation continued until the Industrial Revolution; in the nineteenth century, the invention of the steam train meant that for the first time people could travel longer distances.

Business traffic, however, was a common phenomenon; wandering merchants travelled everywhere in the Empire. Business traffic also comprised tours of duty. It should be stated that this kind of travelling was limited to a small rank of officials, belonging to the upper class: governors and authorities of cities and provinces, who also travelled for pleasure and holidays. It is these voyages in particular that are the most documented in literature.

Commuter traffic

As stated, suburbs as we have nowadays did not exist in the Roman Empire. By far the most employment was situated in the cities; people lived where they worked. Shops and workshops were on street level; the house was behind or above. In fact, in such cities there was no commuter traffic; the shopping public created traffic.

One exception was caused by certain workshops which created inconvenience. Leaving a city, one passed graves and tombs and straggling ribbon development where workshops were situated, causing noise and fire risk, which were not permitted inside the city. These workshops were usually potteries and glass works. In a densely populated city like Cologne we see such workshops along the exit routes of the city;[360] other cities were less densely populated and such workshops were also situated inside the boundaries, like Nijmegen[361] and Trier.[362] In Cologne, a manufacturing city, there must have been a considerable traffic flow inside and outside the walls. Literary and epigraphical sources say nothing about it, but archaeological sources give more information. Figure 2.1 shows that the main concentrations of workshops causing fire risk and nuisance were situated outside the walls and along the most important exit routes to the north, the west and the south. The presence of raw materials made Cologne a centre of glass works and potteries, as well as for metal, stone and leather workshops. In the first, second and fourth centuries glass works, metal workshops and potteries exported their products.[363]

Cologne had some 40,000 inhabitants in the first centuries AD and we know the names of about 300 workshops;[364] a concentration of potteries situated on the west exit route produced about 100,000 items a year.[365] According to Riedel, in a so-called *terra sigillata* pottery there were around 100 workers.[366] Consequently, there must have been considerable traffic passing the city gates.

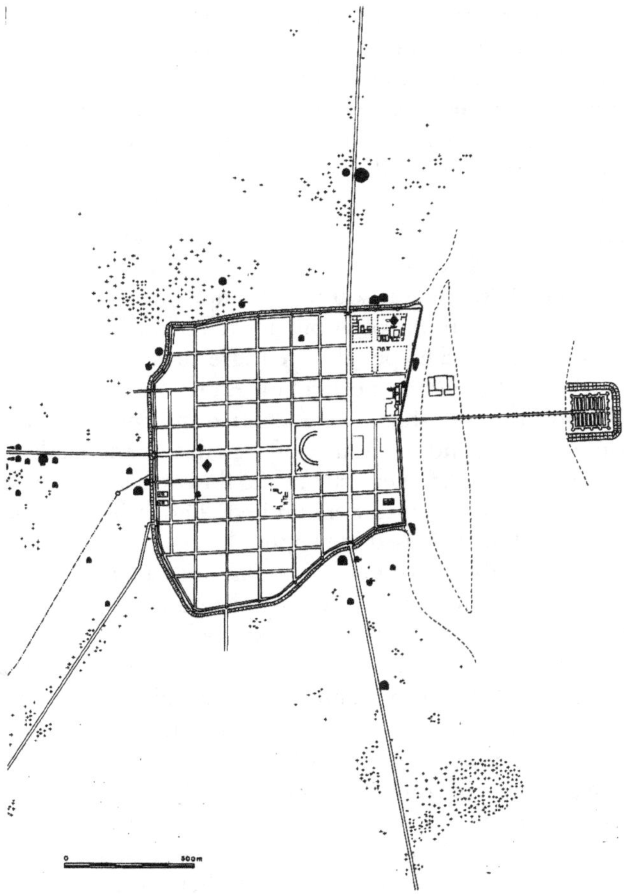

Figure 2.1 Workshops and cemeteries inside and outside the walls of Cologne. Stuart and de Grooth 41. Heerlen: Thermenmuseum.

We can assume that this traffic was mostly pedestrian. The distances to the workshops were often no more than some hundreds of metres[367] and it is improbable that the workers, for the major part slaves, could travel other than on foot. To sustain the traffic flow, the gates had special passages for pedestrians; the main gates to the north, west and south had three passages; another gate on the west side had two passages.[368] The connection between gates and extramural building will be discussed below.

The presence of cemeteries in the neighbourhood of the city walls and exit routes caused smaller-scale traffic consisting of gravediggers and mourners. These people also had to pass the city gates, but they did not cause traffic congestion on a large scale. Funerals took place by day.

Another form of commuter traffic was inside the city walls. Officials and clerks had to travel between their homes and their offices. Most of the cities covered a small area and the distances were short enough for walking. The municipal upper class lived in houses far removed from places where busy goods transport took place, which caused noise and nuisance, but close to the city centre, where the government buildings were situated.

In Xanten, the municipal upper class lived in the west part of the city,[369] and the most important economic activities – trade and industry – took place in the east part, where the harbour was also situated.[370] In Pompeii such a pattern is also visible. The harbour was south of Pompeii and accessable via Porta di Stabia; here the busiest traffic of Pompeii took place.[371] The largest houses of the city, where the richer people lived, are mostly found in the north-west area of the city, although imposing homes are also found in other parts.[372] Because the government offices were situated in other parts of the city than the economic centres, and the houses of the officials and clerks were separated from the houses of the workers, the traffic flow was also, generally, separated; traffic congestion in the cities were limited. The traffic circulation in Xanten and Pompeii will be discussed in chapter 4.

What about commuter traffic in Rome? Rome was not a planned city like Pompeii and Xanten. Instead, Rome had developed into a vast metropolis, without any structure, consisting mostly of working-class quarters. In the centre were the *fora* (business centres) with their temples, law-courts and government centres (Curia, Tabularium) and – in the Empire – the imperial palaces at the Palatine, flanked by the Circus Maximus. North of the Colosseum there was the Carinae quarter, where the imposing houses of important Romans, e.g. Mark Anthony, were situated.[373] Around this centre a vast mass of working-class houses extended, almost completely consisting of slums – with islands, temples, theatres and, from the first century AD onwards, large-scale bath-houses (*thermae*). In fact there were hardly any workshops and the small number of workshops which did exist were set up for the population itself, not for export. In literature concerning ancient economy, Rome is mostly described as a parasite: not producing, only consuming.

The real commuter traffic in Rome was mainly limited to a relatively small group: the politicians with their circle of officials and clerks, who ruled the city and the Empire. More or less comparable with this was traffic caused by clients, on their way to and from their *patroni* for their morning salutation. The population density was higher than in Alexandria and Antioch,[374] and because Rome did not have a good street system, for many it must have been very difficult to reach their office; moreover, the streets were very busy with merchants, shoppers, the unemployed and beggars.[375] Lictors, accompanying the consuls and other politicians, had to eject the crowd, sometimes with violence.[376]

Service traffic

On an interurban level, passenger traffic can be divided into two categories: service traffic (obligatory) and recreational traffic. Besides that, at the interurban level the postal service and the army were active; they will be discussed in the next sections.

Province governors and other (state) officials, including tax officials, travelled frequently throughout the whole Empire, to new provinces or back home. At first, the state took care of transport; food for men and animals was bought by corn dealers (*redemptores*).[377] For lodging, they had to sleep in private houses (*privata hospitia*).[378] Later, when the Empire extended, they took what they needed for their journey. In other words: transport, animals and lodging facilities were merely demanded from the local population. In the period 179–168 BC consul Postumius was the first to demand such a large amount of animals and inns from Praeneste (an ally of Rome) that the local government had to supply the local population with aid.[379] Especially in the Late Republic, when provinces were seen as conquered land which could be plundered as much as possible, many proconsuls and propraetors must have used this as a kind of taxation. From the end of the second century BC onwards the population had, in any case, to supply hay, salt, wood (for cooking), free lodging, draught animals, wagons and ships.[380] Cicero, on his way through Asia Minor, saw that the inhabitants suffered heavily under these forms of taxes.[381] In Italy, the situation was similar.[382] To create a better system, Augustus reorganised the situation and installed a new state transport system, the *cursus publicus*, which could also manage, as well as postal transport, the tours of duty of state officials. The *cursus publicus* was in fact the state transport organisation, transporting everything belonging to the state, people as well as goods. The rule that it was not the voyager but the local inhabitants that had to pay for the journey, food and lodging, was maintained throughout Antiquity.[383]

A person using the *cursus publicus* by order of the state got a ticket (*diploma* or *evectio*, in Late Antiquity also called *tractoria*), giving him the right to use the animals and wagons available. On this *diploma*, the provider's name (with seal), the passenger's name, the available means of transportation (including the animals), the route (including the lodges)[384] and the period of validity were indicated.[385] Travelling with the *cursus publicus* without licence was subject to severe punishments. Perhaps the most well-known example of abuse of the *cursus publicus* was the case of P. Helvius Pertinax. As officer of a cohort, he travelled in Syria by *cursus publicus* without a *diploma* and, after a severe scolding by the governor of Syria, had to walk back from Antioch to his point of departure.[386]

In the light of the abuses of the *cursus publicus* – friends and family of the holders of the *diplomata* come to mind, who were also able to make use of it[387] – we can assume that there were relatively large numbers of travellers with the right to use it. Especially in Late Antiquity, a period with more provinces, more capitals and more bureaucracy, traffic flow increased. Participants of the provincial government meetings (*concilia*) had the right to make use of wagons, but in some cases they had to take care of animals themselves.[388] Bishops and archbishops, participants of ecclesiastical *concilia*, were also entitled to use the *cursus publicus*, which they did. The council of Nicaea in 325 was attended by 250 to 270 bishops, the council of Serdica by 170 and of Rimini by as many as 400. However, not all bishops went by *cursus publicus*.[389] Julian attempted to tackle the abuses, but by way of his anti-Christian policy, ejecting them so often that they were continually *en route* by *cursus publicus* to the next council, in turn obstructing other users of the *cursus publicus*.[390]

Service traffic also includes people with a travelling profession. Besides the postal service, the army and goods transport (traffic flow which will be discussed later in this chapter), there were the travelling merchants, pedlars, wandering philosophers and scientists (geographers like Strabo and orators like Aelius Aristides), sportsmen (especially in the Hellenistic part of the Empire),[391] doctors[392] and artists.[393] Doctors who were not recognised were called *circulatores* and must have had similarities with the later medieval quacks.[394]

Inns

Much is written about inns and places for changing horses, built along the roads, in Antiquity and thereafter. Here is a short description with some general information.

There were two categories of places for changing horses: the *mutationes*, stables where travellers could merely change horses (and other animals), and *mansiones*, inns where travellers could also sleep and eat. *Mutationes* and *mansiones* were known as *stationes*, 'stopping places'.[395]

Mutationes were situated every 10 miles along the road.[396] Between two *mansiones* one could find several *mutationes*; in one day, it was possible to travel from *mansio* to *mansio*.[397] In these *mansiones* were lodging facilities for people and animals;[398] one could eat a (hot) meal and take a bath and a prostitute. A fine example of this can be found on the gravestone of L. Calidius Eroticus (*sic*!). Here follows a short translation: 'Let us pay. Wine and bread 1 *as*, a dessert 2 *asses*, a prostitute 8 *asses*, hay for the mule 2 *asses*. That mule will make me bankrupt!'[399] Much is written about this colourful inn life, especially by Petronius (*Satyrica*) and Apuleius (*Metamorphoses*). We also know about a lot of inns by means of excavations; e.g. that inns were also situated in cities.[400] In Pompeii, we know of the 'Elephant', the inn of Sittius named after its picture of an elephant.[401] In the neighbourhood of city gates, also outside the city, were inns like the 'Elephant', used by the *cursus publicus*.[402]

Although *mansiones* had more accommodation than *mutationes*, they had a bad reputation; many travellers preferred to lodge in private homes or, if possible, friends' homes.[403] Horace makes use of inns and facilities of private homes on his way to Brindisi, but his experiences with inns are bad; he had to put up with bad drinking water, midges, frogs and snoring.[404]

Recreational traffic

In Antiquity there was also recreational traffic. One visited one's family, left the city or went – like today – on holiday. The elite and rich people especially went to their villas and cottages.[405] Places like Tusculum, Tivoli and Baiae were a big attraction for Roman society, passing their time here with pursuits such as gardening, literary debates and – in Baiae – boat trips.

To reach these places one usually made use of the *cisiarii* and *iumentarii*, hirers of wagons and animals, usually located in the neighbourhood of the city gates, e.g. in Verona,[406] in Rome amongst others at Porta Capena and in Ostia at the gate to Rome; in Pompeii they were also located beside the gates (amongst others Porta di Ercolano)[407] between the city boundary and the first milestone.[408] Anyone who did not have permission to use the *cursus publicus* had to turn to these hirers. In cities, the hirers were usually not far away; for this reason, private possession of wagons for passenger purposes must have been scarce. In the countryside, however, the situation was different; to travel anywhere, a farmer or gentleman farmer managing a *villa rustica* or *villa urbana* needed at least one coach or wagon.

Those who could permit it were able to make a long journey to a destination further away. Greece and Egypt especially were favourite destinations. Some Greek travel guides are known describing Greek (mythological) sights: Pausanias' Ἑλλάδος Περιηγήσεως[409] and Heraclides' Περὶ τῶν ἐν τῇ Ἑλλάδι πόλεων. The fact that such guides were written is evidence that there was interest in Greece as a travel destination. Greece was also a favourite province in which to study.

The most usual way from Rome to Greece was the route Via Appia, Via Domitiana or Via Trajana to Brindisi; from there by ship crossing the Adriatic Sea to north-west Greece and along Via Egnatia further to the east.

Another 'holiday destination' was Egypt. Travelling by land was, in many cases, a vast detour, taking a long time. Travelling by sea was shorter and more comfortable and grain-ships with Egypt as destination offered many possibilities; a journey took ten to twelve days.[410] After arrival in Alexandria, a Nile cruise was possible to Memphis, including a visit to the pyramids, Thebes and the colossuses of Memnon.[411]

Of course, the emperors also sometimes made long voyages; for inspection, but sometimes also as tourists. Suetonius mentions that Augustus had visited all provinces, except Sardinia and Africa.[412] Hadrian, known as the most travel-loving emperor, also visited tourist attractions. The travelling emperors will be discussed below, pp. 76–8.

Some people travelled looking for a solution to medical and psychological problems. Everywhere in the Empire, including the western part (especially Gaul),[413] there were facilities where one could be cured of sicknesses. On the Peutinger Map, these cure places are marked by prominent signs. The most important places for convalescing were the sanctuaries of Asclepius: Epidaurus, Pergamum and Cos;[414] the orator Aelius Aristides travelled to Pergamum amongst others for medical advice which will be discussed below. Philosophers also advised trips to cure lovesickness.[415] Seneca mentions that travel was good for a man; discomfort like hunger, cold, a jolting vehicle was good for a healthy mind. For someone too restless, it was advised to stay at home.[416]

Pilgrimages

In the pre-Christian era people were already travelling for religious purposes; the Oracle of Delphi and the Mysteries of Eleusis come to mind.[417] More large-scale pilgrimages began in the fourth century, after the ratification and legalisation of Christianity.

From this century we know of two travel schemes or travel schedules: the *Itinerarium Burdigalense* from 333, where an anonymous pilgrim travels from Bordeaux to Palestine and back again, and the *Itinerarium Egeriae* from the years 381–384, when the female pilgrim Egeria (or Aetheria) makes a pilgrimage to Palestine, Egypt, Mesopotamia, Asia Minor and Constantinople.[418] How long and strenuous such a long voyage was is shown by figures from the *Itinerarium Burdigalense*: from Bordeaux to Constantinople: the pilgrim travelled 4021 miles, making use of 230 *mutationes* and 112 *mansiones*.[419] In spite of the long distances, it was very crowded in Palestine; at the end of the fourth century, Jerome complains about souvenir-selling, the showy make-up used by women and the crowds in Jerusalem.[420] For Jerome, it must have been a source of irritation.

Good and bad travel experiences

A trip could be just as good or bad as nowadays. Apart from that of Horace there are further journey reports; here are two examples.

On 12 April, in a year between 317 and 323, the official Theophanes travelled from Egypt to Antioch and back again. He probably used the *cursus publicus*, because we do not read about the renting of animals or lodges; however, he had to provide himself with food and drinks. Many of his companions were slaves and there was also a clerk. On 30 April, eighteen days after his departure, he reached Antioch; on 19 July he started on his journey home and on 5 August he arrived back in Egypt again.

On his departure, Theophanes bought plenty of food: fine bread for himself and his table-companions, cheaper bread for the slaves, barrels of wine, meat, grapes, apricots, melons, cabbages, olive oil, fish sauce, honey, fire-wood, apples and sausages. On his way the following products were bought: grapes, figs, peaches, apricots, toilet articles, soap, eggs, veal, two kinds of wine (one for Theophanes and one for the slaves), fresh meat (slaughtered at a *mutatio*), etc. Besides that he had, just like every traveller starting a long journey, bedding and cooking utensils.

The fact that he needed a lot of luggage shows that Theophanes travelled with many companions and that a long convoy was under way. Unfortunately it's not clear if he travelled with pack animals or wagons; maybe he used both. On his way to Antioch, the route Laodicea–Antioch (more than 64 miles) was travelled in one single day; we can assume that it was a good trip.[421]

The fact that a journey was not always a good one[422] is shown by the reports of the aforementioned sophist and orator Aelius Aristides. In 165 AD he travelled from Smyrna to Pergamum, to recover from illness. In the beginning he also travelled with a wagon train, but he sent it onwards. He hoped to reach Myrina, but he did not arrive earlier than midnight, after discovering that in Laryssa and Cyme everything was closed. He finally spent the night in the house of a friend. The next day he sacrificed to Apollo and the day after he finally arrived in Pergamum.

Aelius Aristides had complaints, not only with his health, but also with the discomfort during the journey. He suffered from heat, cold, blowing sand and closed inns, so that he had to travel by night in order to find an inn which was open; which he did not find.[423] Probably, Aelius Aristides did not travel by *cursus publicus*; he was not an official and did not sleep in prescribed inns. He regularly mentions tired slaves; they probably had to travel by less comfortable means than Aelius Aristides himself, for instance in a simpler wagon or mounted on animals.[424]

Finding the way

Another story is how to find the way in a (strange) city. We know the names of the most important streets of Rome like Via Appia, Alta Semita or Argiletum. But we know neither the innumerable names of all the lanes and cul-de-sacs where the majority of the inhabitants had to live or make their living, nor street names in other cities. We do not even know if they had names.[425] In the second century BC, Terence wrote in his work *Adelphi* the following passage:

SYR. *nostin porticum apud macellum hac deorsum?*
DEM. *quid ni noverim?*

SYR. *praeterito hac recta platea sursum: ubi eo veneris, Clivos deorsum vorsum est: hac te praecipitato. postea est ad hanc manum sacellum: ibi angiportum propter est.*
DEM. *quodnam?*
SYR. *illi ubi etiam caprificus magna est.*
DEM. *novi.*
SYR. *hac pergito.*
DEM. *id quidem angiportum non est pervium.*
SYR. *verum hercle. vah, censen hominem me esse? erravi: in porticum rursum redi: sane hac multo propius ibis et minor est erratio. scin Cratini huius ditis aedis?*
DEM. *scio.*
SYR. *ubi eas praeterieris, ad sinistram hac recta platea; ubi ad Dianae veneris, ito ad dextram. prius quam ad portam venias, apud ipsum lacum est pistrilla, ei exadvorsum fabrica: ibist.*

SYR. Do you know the colonnade by the meat-market, down that way?
DEM. Of course I do.
SYR. Go that way straight up the street. When you get there the Slope is right down in front of you: down it you go. At the end there's a chapel on this side. Just by the side of it there's an alley.
DEM. Which?
SYR. That where the great wild-fig-tree is.
DEM. I know it.
SYR. Take that way.
DEM. That's a blind alley.
SYR. So it is, by Jove. Tut, tut, you must think me a fool. I made a mistake. Come back to the colonnade: yes, yes, that's a much nearer way and much less chance of missing it. Do you know Cratinus's house, the millionaire man there?
DEM. Yes.
SYR. When you are past it turn to your left, go straight along the street and when you come to the Church [the Temple of Diana] turn to the right. Before you come to the towngate, close by the pool there's a baker's shop and opposite it a workshop. That's where he is.

(Ter. *Ad.* 573–84)

This is a comedy, situated in Athens, or anyhow a Greek setting, but such a route description could mean that places were very difficult to find – a comic and recognisable situation for the comedy's audience. Street name plates are found nowhere, in any case not as our street name plates. In Pompeii, some tiles have been found functioning as street name plates, with illustrations of a rosette, an ass, flower petals or a phallus.[426] Probably such tiles were also present in other cities, but we have to assume that street names were more the exception than the rule, especially in Mediterranean provinces. In northern Europe street names are unknown.

Another passage, from Petronius' *Satyricon*, tells us about three men, finding their way back at night because one of them had marked the route with chalk:

> *Neque fax ulla in praesidio erat, quae iter aperiret errantibus, nec silentium noctis iam mediae promittebat occurrentium lumen. Accedebat huc ebrietas et imprudentia locorum etiam interdiu obfutura. Itaque cum hora paene tota per omnes scrupos gastrarumque eminentium fragmenta traxissemus cruentos pedes, tandem expliciti acumine Gitonis sumus. Prudens enim [pridie], cum luce etiam clara timeret errorem, omnes pilas columnasque notaverat creta, quae lineamenta evicerunt spississimam noctem et notabili candore ostenderunt errantibus viam.*
>
> There was no guiding torch to show us the way as we wandered; it was now midnight, and the silence gave us no prospect of meeting anyone with a light. Moreover we were drunk, and our ignorance of the quarter would have puzzled us even in the daytime. So after dragging our bleeding feet nearly a whole hour over the flints and broken pots which lay out in the road, we were at last put straight by Giton's cleverness. The careful child had been afraid of losing his way even in broad daylight, and had marked all the posts and columns with chalk; these lines shone through the blackest night, and their brilliant whiteness directed our lost footsteps.
>
> (Petr. 79)

This is a reference to the fact that finding the way in a strange city was not easy, especially not for an individual traveller, and especially at night (street-lighting was unusual) – although neither is the *Satyricon* based on daily reality. The absence of orientation marks like street name plates can also be evidence that the man in the street did not travel on his own. Finding the way could be left to the carter or coachman.[427] People travelling on their own with a team of slaves could send a slave on to explore the city, like the aforementioned Aelius Aristides; a user of the *cursus publicus* had no problems in finding his way.

Animals and wagons

The animal most used was the ass or mule.[428] These animals had some advantages over the horse: they were cheaper, and had more stamina and harder hooves.[429] The authors Varro and Columella describe the advantages of the ass;[430] Pliny the Elder describes those of the mule.[431] The fact that one still comes across asses nowadays in the countryside of Greece and other Mediterranean countries shows their advantages.

For speed, horses were used, especially for drawing two-wheeled vehicles. In Late Antiquity, the horse was more attractive as a riding animal with the introduction of saddle and stirrups; at this time the introduction of the horseshoe also took place.[432] However, horses were no higher than ponies are nowadays.[433] Horse riding had some advantages over riding by wagon: faster, more comfort (bad road surfaces!) and fewer holdups.[434]

For slower transportation (goods traffic) oxen were used, which could not move fast, but were strong and had a lot of stamina. A fourth animal, used in the deserts of the East and African provinces as a riding and pack animal, was the dromedary, also known for its stamina.[435]

The *iumentarii* hired riding, draught and pack animals; the *cisiarii* rented wagons. Here follows a short description of the most important vehicles used for passenger traffic.[436]

The *carpentum*[437] was a hooded cart with two wheels, usually richly ornamented. We see the *carpentum* as a wagon not only for ceremonies[438] but also for journeys.[439] A *carpentum* was usually drawn by mules (Fig. 2.2).

A *carruca*[440] was a wagon with four wheels, drawn by two horses or two mules (Fig.2.3).[441] There is disagreement as to whether *carrucae* were open or closed. According to Isidorus, they were closed;[442] this was the case for the most luxurious form of the *carruca*, the *carruca dormitoria* or sleeping wagon; of *carrucae* this type are exhibited in the archaeological education centre Archeon in Alphen a/d Rijn (Netherlands) (Fig. 2.4) and in the Römisch-Germanisches Museum in Cologne. This vehicle had a relatively comfortable interior and was even fitted with springs.[443]

Figure 2.2 Carpentum. Saglio 927. Paris: Hachette.

Figure 2.3 Carruca. Saglio 928. Paris: Hachette.

Figure 2.4 Carruca, Archeon. Photo: C.R. van Tilburg.

The *carruca* in Alphen a/d Rijn, on a par with the wagon in Cologne, has an axle width, including the wheels, of 1.80 m and including the axle heads of 2.20 m; the length excluding the pole is 3.10 m, including the pole 4 m, and the height is 2.35 m.[444] The *carruca* in Cologne has an axle width, measured between the centres of the rims, of 1.60 m.[445] In contrast to the *carruca* in Cologne, the one in Alphen a/d Rijn has open windows.

The axle width, as well as the width of the wheels, could vary. Grénier makes a distinction of three gauge widths: 'narrow gauge' from 1.10 to 1.20 m and sometimes narrower, suitable for mountain roads (narrow wagons could, of course, also be used on flat land);[446] 'normal gauge' from 1.35 to 1.45 m; and 'wide gauge' from 1.55 to 1.65 m and wider.[447] The aforementioned *carrucae* in Alphen a/d Rijn and Cologne had axles of the last category. In Pompeii and its surroundings wagons have been found with axles of 'broad gauge' and 'normal gauge', varying from 1.42 m to 1.58 m.[448]

The *cisium*[449] was a plain two-wheeled open light-weight wagon (Fig. 2.5). This was very attractive in view of its high speed and it was drawn by one or two animals (horses or mules). The *birota* or *birotum* was also used for the *cursus publicus*: a light two-wheeled wagon, only for passenger use. It must have looked, generally, like the *cisium*.[450]

The *rheda, reda, raeda*[451] was a heavy four-wheeled wagon, usually drawn by eight to ten horses or mules.[452] It was used for heavy cargo and several passengers; very suitable for family travel. It has already been mentioned that the *cursus publicus* used it for passenger transport.[453] It was a plain open wagon without much luxury. A rented wagon was called *reda meritoria* and one could rent it from a *raedarius*. In Cicero's *Pro Milone* there is talk of a robbery where the *raedarius* is killed; one can conclude that a *raedarius* could have been a coachman or carter.[454]

Isidorus mentions other types of wagons in his time, like the *caracutium*, the *pilentum* and the *basterna*.[455] Besides these, he also gives attention to another means of conveyance: the sedan. Sedans also could be divided into

Figure 2.5 Cisium. Saglio 1201. Paris: Hachette.

several categories and had the advantage of being allowed to go everywhere, especially in narrower roads and streets.[456]

Costs

Finally, some information on the rental costs and purchase prices of wagons and travelling. At the end of the third century, Diocletian gave a mandate for a maximum price of goods and services in the so-called *Edictum Diocletiani de pretiis*.

A *rheda* with *arcuatae rotae* without iron mounting cost 3000 denaries; a sleeping-wagon with *arcuatae rotae* without iron mounting 4000 denaries; a sleeping-wagon with *vitutae rotae* without iron mounting 7500 denaries; and a *carruca* with *vitutae rotae* without iron mounting 7000 denaries.[457] *Rotae vitutae* were one-piece wheels with one round rim (*vitus*, 'rim'); *rotae arcuatae* were wheels which were constructed from several pieces (*arcus*, 'round pieces').[458] If one compares the remarkable differences in price one can assume that wagons with *rotae vitutae* were more comfortable and the wheels were perhaps more durable. The *carruca* mentioned here was probably an open one. Iron mounting was not included; that had to be done by a coachbuilder or a smith.

It was much cheaper to rent a wagon with a coachman, if the distance was not too far. The rent price for a *rheda* was 2 denaries per person per mile and for cargo by *rheda* 12 denaries per mile.[459] Unfortunately, the prices of riding and draught animals are not known, but the price of food for pack animals is documented. Goods transport will be discussed below, pp. 68–76.

Summary and conclusion

Compared with our traffic nowadays, passenger traffic in Roman times by wheeled vehicle was scarce. Without doubt the most important passenger traffic flow was pedestrian. The distances for commuter travel were short; one lived in the same city, village or settlement where one had one's job. Province governors and state officials travelled longer distances (paid by the state) from province to province. Other people, able to pay for themselves, travelled for recreation, study or pilgrimage.

The physical aspect was an important factor. In spite of the well-paved roads, the road signs (milestones) and the many resting places along the roads, travelling was an exhausting matter, even in the most comfortable wagons and coaches. Hunger, thirst, cold, heat and crime were big problems. Those who had enough money or a licence to travel by *cursus publicus* had more chance of having a good journey than someone who had to travel on his own. Travelling by ship had more advantages, but not all places were approachable by water and in wintertime there was no water traffic because of ice and storms.

It is impossible to quantify passenger traffic flow, even to make an estimate. Only incidental travel descriptions of (wealthy) travellers give us a vague idea of their travelling-companions and their wagons and animals. Using this information, we can conclude that these people did not usually travel without a crowd of slaves. Slaves were probably not even an extreme luxury during travel. They could give protection against banditry, could provide travel requisites and could find the way in a strange city.

Postal service and *cursus publicus*

Introduction

A very important part in the functioning of the Roman Empire was the existence of the state transport organisation, the *cursus publicus*, transporting passengers and goods as well as the post. The need of an optimum working message service was already acknowledged at an early stage and created an important reason for building an extensive and well-paved road-system.

In the Empire especially the *cursus publicus* was an important road-user. One can say that employees and officials of the *cursus publicus* were intensive road-users and much of the accommodation along the roads, especially the *mansiones* and *mutationes*, was based on the needs of the *cursus publicus*. With the passing of time, this organisation became not only more and more important, but also a growing problem for the population. The abuses of the *cursus publicus* have already been mentioned in the previous section; in this section this aspect will be elaborated upon.

The *cursus publicus* used three forms of transport: pedestrians, horsemen and wagons. At first, pedestrians formed the main category; later, in the extended Empire, the importance of fast transportation was pre-eminent.

What is striking in all publications concerning the *cursus publicus* (especially the work of A. Kolb) is the large amount of officials' names, especially in Late Antiquity when the *cursus publicus* was in the period of enlarged bureaucracy; our most important written source for the *cursus publicus* is the *Codex Theodosianus*, dating from the fifth century. In the interests of clarity, I shall not reference these in this book.

Post in the Republic

Everyone, even the writer himself, could deliver a letter or message. When Rome was still a minor city, there was no need for fast communication and a paved road-system. Governors like Cicero (governor in Cilicia in 50–51 BC) used soldiers (*statores*) assigned to him to send official messages; besides that, lictors were also used.[460] However, governors also used professional couriers, state slaves (*viatores*), for their official correspondence.[461] For their private correspondence they used their own slaves or slaves of the addressee, in the same way that Cicero used the service of Atticus' slave Nicanor.[462] Couriers bringing written messages were, generally, called *tabellarii*; if they brought oral messages, *nuntii*.[463] Further there were *tabellarii publicanorum*, travelling tax officials also entitled to deliver letters.[464]

The majority of the messengers were pedestrians and they were known as *cursores*, 'runners'. Only in a few cases, with the message of a victory or a defeat when speed was important, could they use horses (messengers on horseback were known as *veredarii*) and *cisia*. In Republican times, there were already *cisiarii* in Praeneste.[465]

During the Republic, the relay system had already been introduced; messages and/or letters were handed on to several couriers. In this way, Caesar's victory at Pharsalus was proclaimed.[466] During the civil wars especially the importance of fast communication was recognised. Nevertheless, post was not completely secure. Cicero complains that not all letters reached him and other letters came too late, amongst others things because of robbery.[467] Most victims of robbery would have been the pedestrian couriers.

The introduction of the cursus publicus

To make communication as good as possible, Augustus founded a state transport company, conveying all state property: officials, goods, money, etc. Later, this company was named *cursus publicus*, but before that it was known as *vehiculatio*; the name *cursus publicus* is first mentioned in the fourth century AD.[468] Exactly when the *cursus publicus* was founded is not known; Suetonius describes that Augustus soon took measures after the beginning of his monarchy to obtain fast communications by means of young relay runners and couriers by wagon. The advantage of couriers by wagon was that they did not have a relay function and they could have charge of the whole message during the journey and pass on the information in person.[469]

Although the *cursus publicus* for the most part was meant for transport of officials and goods belonging to the state, sometimes couriers could also use it. They could use the lodging facilities along the roads, but did not change animals. For urgent messages, however, they could, in accordance with the relay system, use fresh riding animals, probably horses. The *cursus publicus* did not have its own couriers.[470]

Passenger transport has already been mentioned. Now some general aspects of the *cursus publicus* and goods transportation will be quoted.

The praefectus vehiculorum

The *cursus publicus* was administered by the *praefectus vehiculorum*. It is not exactly known when this function was introduced. In the first century

AD some *praefecti vehiculorum* are known, and because Augustus introduced several functions with the name *praefecti* (amongst others the *praefectus vigilum* and the *praefectus annonae*) we can assume that, with the introduction of the *cursus publicus*, he also inaugurated this function.[471]

The *praefecti vehiculorum* – coming from the ranks of the knights – were only active in Italy at first; in the provinces, the governors were responsible for the smooth functioning of the *cursus publicus*.[472] Names of Italian *praefecti vehiculorum* include Baebius Juncinus (under Nero)[473] and Q. Marcius Turbo (*c.* 110 AD).[474] At the end of the second century and in the third century, we meet them in some provinces, e.g. Q. Julius Maximus Demetrianus (beginning of the third century) in Belgica and the Germanic provinces[475] and one unnamed (268 or later) in Gaul.[476] Also dating from the third century onwards is the appointment of *praefecti vehiculorum* for individual roads, e.g. Via Flaminia.[477] This kind of administration and organisation shows similarities with the *curatores viarum*, probably also only active in Italy and each of them was responsible for only one road. During the reign of Constantine, there were still *praefecti vehiculorum*, but it is unknown where they were active; as postal manager we know of a *praepositus de via Flabinia* (*sic*).[478] The *praefecti vehiculorum* were probably placed directly under the emperor.[479] We do not know much about subsidiaries of the *praefecti vehiculorum*. During Gallienus' reign, we know of the *advocatus fisci* A. Vitellius Felix Honoratus;[480] further a freedman *tabularius a vehiculis*[481] and an *acceptor vehiculorum* Flavius Aug. lib. Pergamus.[482]

Another task of the *cursus publicus* was, originally, the provisioning of the army; for that reason the *praefectus vehiculorum* was, in the first instance, also responsible for this task. Such a man was Plotius Grypus, who had this function under Domitian.[483] Besides that, he supervised the functioning of the *mansiones* and *mutationes*, administered by *mancipes*.[484]

Goods transport by **cursus publicus***:* **cursus velox** *and* **cursus clabularis**

The *cursus publicus* also transported goods; however, goods transport was less important than passenger transport.[485] An edict of the governor of Galatia shows that travelling officials had permission to take goods with them, on a modest scale.[486] During the reign of Septimius Severus[487] there were two categories of the *cursus publicus*: the *cursus velox* for fast post, not occupying much space, and the *cursus clabularis* or *clavularis*. These names were first used in Late Antiquity, in written documents.[488]

The *cursus velox* transported goods demanding high speed, e.g. tax money and metals such as gold and silver, with high priority. For this kind of transport horses, asses and mules were used.[489] The post-horses were called *veredi*, *βερέδοι*.[490]

The *cursus clabularis* transported goods where a fast service was not so urgent and transported more weight and volume, e.g. marble[491] and army equipment.[492] The *cursus clabularis* made use of oxen; the wagons were the aforementioned *rhedae* and (for passenger transport also) then used *angariae*, heavy wagons.[493] In the *Codex Theodosianus* the *carrus* is also mentioned, like the *rheda* first equipped with two wheels and later with four; with this wagon passengers and goods could be transported, but it seems to have been mainly a cargo wagon.[494]

The goods transport of the *cursus publicus* was subject to a maximum weight (Table 2.1). For the *rheda*, there was a maximum weight of 1000 pounds (326 kilo);[495] for the *sacrae largitiones* it was permitted to carry 550 pounds of gold (500 on the wagon itself and 50 by the accompanying slaves) or 1000 pounds of silver; for the emperor's private property the maximum weight was 300 pounds of gold or 500 pounds of silver.[496]

There was strict control of the maximum weight. A law enacted by the emperors Valentinian I, Valens and Gratian made it possible to erect special checkpoints to control the maximum weights of *angariae, rhedae* and horses.[497] Wagon-builders were punished with exile or penal servitude in the mines when they built larger wagons.[498] The purpose of this severe legislation concerning maximum weight was to keep the costs of the *cursus publicus* under control; one wanted to spare the animals (which local inhabitants were forced to supply).

The cursus publicus *and the charges to the population*

It has been explained above (p. 45) that even before the introduction of the *cursus publicus* the local people could be forced to supply free animals, wagons, food and lodgings to important passengers whilst on their way. In fact, the introduction of the *cursus publicus* did not change anything; the obligations were just better organised.

These obligations were known as *angariae*, a Persian loan word introduced via the Greek word ἀγγαρεία. The first mention of the Latin word *angariae* dates from the year 159/160 AD in Moesia and is in fact a complaint.[499] Besides these *angariae* there were other obligatory services and taxes (λ(ε)ιτουργίαι, *munera*) all to do with the *cursus publicus*. How much these *angariae* oppressed the people is demonstrated by the fact that several emperors tried to alleviate it. An inscription from 49/50 AD makes it clear that Claudius tried to moderate the tax, but failed.[500] Vespasian, reigning with a money-reducing policy, did not spare the people with his high taxes.[501]

The first emperor taking effective action was Nerva. Italy, by reason of the dense road-system and the large population, which without doubt suffered the most under the obligations, was freed from the *angariae*, as shown on a coin (Fig. 2.6) with the inscription *Vehiculatione Italiae remissa*.[502]

Table 2.1 Maximum weight of the *cursus publicus* (goods transport)

Angaria	1500 pounds[a]
Rheda (general)	1000 pounds[b]
Rheda (gold for the *sacrae largitiones*)	500 pounds
Rheda (gold for the *res privatae*)	300 pounds
Rheda (silver for the *sacrae largitiones*)	1000 pounds
Rheda (silver for the *res privatae*)	500 pounds[c]
Carpentum	1000 pounds and 2 to 3 escorts[d]
Carrus	600 pounds[e]
Birota	200 pounds[f]
Pack-horse	100 pounds[g]
Horseman	30–35 pounds[h]

[a] *Codex Theodosianus* 8.5.28 *angariae mille quingenta sufficiant.*

[b] *Codex Theodosianus* 8.5.8.1 *Statuimus raedae mille pondo tantummodo superponi*; 8.5.28 *non amplius raeda quam mille pondo subvectet.*

[c] *Codex Theodosianus* 8.5.48.pr. *si aurum sacrarum largitionum vel argentum ad comitatum nostrum destinatur, una raeda quingentis auri lib(ris), mille vero argenti, si vero privatarum, auri trecentis, quingentis vero argenti libris oneretur.*

[d] *Codex Theodosianus* 8.5.30 *sanxeramus, ut in carpentis raedarum mensuram subditam nullus excederet*; 8.5.18.1 *ne amplius in singulis quibusque carpentis, quam bini aut ut summum terni homines invehantur, quos tamen directarum rerum custodes vel prosecutores esse constiterit*; Hudemann 151.

[e] *Codex Theodosianus* 8.5.47.pr. *carro sescentarum nec amplius addito eo.*

[f] *Codex Theodosianus* 8.5.8.1 *Statuimus* [. . .] *superponi, birotae ducenta.*

[g] Cassiodorus, *Variae* 4.47.5 *Parhipporum quin etiam onera centum librarum nullus excedat*; A. Kolb 2000, 216.

[h] *Codex Theodosianus* 8.5.47.1 *veredorum* [. . .] *sexaginta libras sella cum frenis, triginta quinque vero averta non transeat*; A. Kolb 2000, 216 ns. 6 and 7.

Nerva's successor Trajan reversed this measure for the most part and continued the former policy, but he had more control over the allocation of *diplomata*; before that time, there was a very liberal allocation. Trajan's policy was also not always consistent; Pliny the Younger asks permission retrospectively for such a *diploma* for his wife, travelling by reason of bereavement. He gets it.[503]

Later emperors regularly tried to alleviate the situation by allowing certain groups of professions and/or regions to be free of the charges of the *cursus publicus*. During the reign of the Severan emperors, soldiers and veterans travelled free.[504] When in the third century the political instability increased, the activities of the *cursus publicus* also had to increase: more troop movements, more mobility of the imperial court and, inevitably, a bureaucracy who travelled more. Combined with a decreasing population, the tax burden must have been intolerable for vast sections of the Roman Empire.

Figure 2.6 Coin of the emperor Nerva. Teitler 132. München: Hirmer Verlag.

Constantine demanded in 315 that users of the *cursus publicus* had to wait until an ox of the *cursus publicus* arrived if there were none available; he also ordered that the use of farmers' oxen was forbidden.[505] In fact, the entire eighth book of the *Codex Theodosianus* is a description of laws against extreme abuses of the *cursus publicus*. Cases like cruelty to animals,[506] the huge supply of tickets[507] and the misuse of company facilities for private purposes,[508] forbidden according to this *Codex*, must have been normal phenomena. Because of cruelty to animals, the *cursus publicus* continually needed more animals and this caused a shortage.[509] Julian decreased the distances between the stopping places to spare the animals,[510] he forbade the use of more animals than necessary[511] and in Sardinia the *cursus velox* was abandoned, because there was no need for fast communication on horseback.[512]

The fact that – in Late Antiquity especially – the *cursus publicus* was used not only as a means for communication and postal service but also as a way of controlling local inhabitants by the so-called *agentes in rebus* must have increased the community's hatred of the company.[513]

The volume of traffic flow: rough estimations

In the *Notitia Dignitatum*, an enumeration is given of the allocation of tickets to several high state officials. In total, there are 182 tickets provided annually.[514] Should we estimate that goods transport and illegal use doubled this figure, we have some 550 journeys with the *cursus publicus* each year. The *angariae* may have severely oppressed the community, and the transportation of high state officials by *cursus publicus* must have played a modest role in the whole logistic. If we assume the number 182, we see that, on average, one journey took place every two days by *cursus publicus* in the east part of the Empire. If we assume the number of 550 journeys, we see one-and-a-half to two journeys each day. An important official, travelling with many companions, could however cause a considerable jam among other traffic.[515]

The number of animals in a *statio* can also give us information on the volume of the *cursus publicus*. Hudemann states that a *mutatio* could contain twenty horses and a *mansio* forty or more. Perhaps these numbers are based on Procopius, who mentions the number of forty animals.[516] A. Kolb and Stoffel, however, dispute this number. According to the *Codex Theodosianus*, it was forbidden to set more than five to six horses before a *rheda*; in Justinianus' time, this number was ten.[517] Kolb estimates the total number of animals per *statio* as between nine (four horses, five mules) and twenty (ten horses, ten mules), with two to four oxen; in total a maximum number of twenty-four animals, but probably more per *statio*.[518] There is at any rate a case of an official using thirty asses and ten horses for a journey.[519] In Tiberius' time, there is an inscription showing that a village had to supply ten wagons, thirty mules or sixty asses.[520] Another inscription gives an enumeration of the maximum number of wagons and animals that could be used, depending on a person's rank. A senator could use ten wagons, a knight three and a *centurio* one.[521]

When we compare this large number of animals with the relatively small portion of the *cursus publicus*, it is probable that other road-users were also able to use these animals, mostly for money, like the *agentes in rebus* (there were more than 1000 of them),[522] the private *iumentarii* and probably people who travelled on their own and/or transported goods.

Summary and conclusion

For the Roman Empire to be managed as well as possible, a fast communication system and well-organised goods transport were necessary. Originally, slaves and soldiers took care of the sending of messages. Later this task – in so far as it belonged to the state – was executed by the *cursus publicus*, founded by Augustus. This *cursus publicus* also transported officials, gold, silver, money, marble and army equipment.

It is typical of the *cursus publicus* that its users, travelling as holders of a *diploma* or *evectio*, did not pay (or did so only partially) for travelling. The local population was forced to supply everything that they needed: animals, wagons and food. These supplies, called *angariae*, were burdensome and several emperors tried to alleviate this. In the fourth century especially, when the army and the bureaucracy increased and thereby also the use of the *cursus publicus*, strict laws were needed to protect the population against large-scale abuse of the *cursus publicus*.

Army and road security

Introduction

The Roman army, one of the main groups for which the road-system was built, was, just as the *cursus publicus*, a relatively small road-user. Most military activities took place in the frontier regions. Particularly during the Empire, legions marched from one frontier province to another; it was very important not to meet many obstacles on their way. This was more important in the case of campaigns, where speed was crucial. The army did not depend on *stationes* along the road, but had their own facilities for food and lodging.

In this section attention will also be given to the security of the roads. Along the whole road-system there were *stationarii* and *beneficiarii*, not only fighting against highwaymen, but also protecting tax officials everywhere in the Roman Empire.

The size of the Roman army

The volume of the Roman army cannot be estimated exactly and an extensive investigation into this is not the subject of this book. However, some statistics are known and indicate the following.

At the beginning of the Republic – 510 BC – Rome had two legions, each numbering 4200 foot-soldiers and 300 mounted soldiers and each commanded by a consul.[523] In the fourth century BC, the division into maniples took place: a legion contained ten maniples with two *centuriae* each. Polybius mentions that a legion comprises 1200 *hastati*, 1200 *principes*, 600 *triarii* and 1000 *velites*. This division was based on age. Sometimes a legion had more soldiers; the proportions of the soldier groups were always the same, except the *triarii* which always had the same number.[524] At the beginning of the Second Punic War (218 BC) the Roman army must have had 10,000 men. Before the Battle of Cannae there were 70,000 Roman soldiers; after this battle, 50,000 had been slaughtered.[525] From this period onwards the army was replenished by *auxilia*, light-armed men and cavalry.[526]

When Marius abolished the capital demand, *c.* 100 BC, and reorganised the army into a larger professional army, the new division of the legions took place. A legion now comprised 6200 men, supplied with 300 cavalry.[527] The division was as follows: ten cohorts of three maniples each, both with two *centuriae*; one cohort also comprised six *centuriae*.[528]

The more the Empire extended, the more the need for soldiers. When Augustus began his reign, there were seventy-five legions, altogether numbering

300,000–350,000 men; during his reign two thirds of them were demobilised, so that twenty-five legions remained. This resulted in a smaller army and large-scale land expropriations. Later the number of legions increased again, to thirty-three in the time of Septimius Severus.[529] An overview of the aforementioned twenty-five legions is given by Tacitus:

> *Sed praecipuum robur Rhenum iuxta, commune in Germanos Gallosque subsidium, octo legiones erant.*[530] *Hispaniae recens perdomitae tribus habebantur. Mauros Iuba rex acceperat donum populi Romani. Cetera Africae per duas legiones parique numero Aegyptus, dehinc initio ab Syriae usque ad flumen Euphraten, quantum ingenti terrarum sinu ambitur, quattuor legionibus coercita, accolis Hibero Albanoque et aliis regibus qui magnitudine nostra proteguntur adversum externa imperia. Et Thraeciam Rhoemetalces ac liberi Cotyis, ripamque Danuvii legionum duae in Pannonia, duae in Moesia attinebant, totidem apud Delmatiam locatis, quae positu regionis a tergo illis, ac si repentinum auxilium Italia posceret, haud procul accirentur.*
>
> Our main strength, however, lay on the Rhine – eight legions ready to cope indifferently with the German or the Gaul. The Spains, finally subdued not long before, were kept by three. Mauretania, by the national gift, had been transferred to King Juba. Two legions held down the remainder of Africa; a similar number, Egypt: then, from the Syrian marches right up to the Euphrates, four sufficed for the territories enclosed in that enormous reach of ground; while, on the borders, the Iberian, the Albanian, and other monarchs, were secured against alien power by the might of Rome. Thrace was held by Rhoemetalces and the sons of Cotys; the Danube bank by two legions in Pannonia and two in Moesia; two more being posted in Dalmatia, geographically to the rear of the other four, and within easy call, should Italy claim sudden assistance.
>
> (Tac. *Ann.* 4.5)

This calculation only concerns the legions. The floating cohorts of auxiliaries, estimated by Clauss at 150,000–200,000 men in total,[531] are not included, nor are the praetorians and city cohorts – which Tacitus calculates after the aforementioned amount – responsible for public security and fire watch.

In Late Antiquity, the army was extended and again reorganised owing to increasing pressure on the frontiers. Now the army was divided into *comitatenses* (field troops) and *limitanei* (frontier troops). The division is shown in Table 2.2.[532]

Together with the imperial guards, the total sum is more than half a million military around 400 AD.[533] The maximum number of soldiers was,

Table 2.2 Comitatenses and *limitanei*

Comitatenses	*Limitanei* (east part)
25 *legiones palatinae* (each 1000 men)	41 *legiones* (each 1000 men)
69 *legiones comitatenses* (each 1000 men)	44 *auxilia* infantry (each 500 men)
38 *legiones pseudocomitatenses* (each 1000 men)	105 cohorts infantry (each 500 men)
vexillationes cavalry (each 500 men)	47 *cunei* cavalry (barbarians)[a]
auxilia up to 500 men (barbarians)	121 cohorts cavalry (each 500 men)
	65 *alae* cavalry (each 500 men)[b]

[a] Liebenam 1623.
[b] In the first and second centuries AD an *ala* comprised a quantity of 500 soldiers, in some cases 1000; Webster 145. I presume that this number was also valid in Late Antiquity.

according to Agathias (sixth century), as much as 645,000, in the second part of the fourth century.[534]

The division of a legion

Under Marius, the divisions of a legion typical of the Empire were introduced: more than 6000 infantry and 300 cavalry.[535] Vegetius, living *c.* 400 AD, gives other numbers in his book *Epitoma rei militaris*: 730 cavalry, and 6100 infantry.[536] However, it is not clear which period he is writing about; in his time (fifth century AD) the division of a legion was totally different from that during the Principate and had yet other numbers.

A *centuria* comprised eighty men; Hyginus mentions *papiliones* of eight men each, so a *centuria* numbered ten *papiliones.*[537] In the *castra* and *castella* along the frontiers, one block contained *contubernia*-lodges with at the head the lodge of the *centurio.*[538] Six *centuriae* made up a cohort, i.e. 480 men; ten cohorts made up a legion, 4800 men. These are the numbers of the Early Empire; it has already been mentioned above that not all legions were of the same size. Moreover, a legion also contained smiths, wagon-builders, carpenters, slaves, etc.[539] Besides men, there were also pack, draught and riding animals; Junkelmann assumes that there were two pack animals (probably asses or mules) available for each *contubernium*;[540] in a complete legion (6000 men) there were 1400 mules and 300 horses, including riding animals for officers.[541] He estimates the total sum of animals in military service at 200,000 in the middle of the second century AD.[542] This number must have been increased in Late Antiquity, when cavalry played a more important role in the defence of the Empire.

An army on its way

There are some descriptions of the division of an army on the march. Flavius Josephus and Arrianus describe that an army (against the Jews and the Alans respectively) had a vanguard of light-armed auxiliaries (including bowmen); thereafter units of cavalry and infantry; then, in the centre of the column, the staff including the *imperator* (general); thereafter the legions, more auxiliaries and cavalry. Luggage and artillery (catapults) were positioned in the centre of the column.[543] This placing of auxiliaries and cavalry in the vanguard and rearguard with legions in the centre agrees with Tacitus' description of a campaign in Germania.[544]

Flavius Josephus states that the legion soldiers marched six abreast; Arrianus mentions the number of four.[545] If we assume that a road had an average width of 18 feet (about 6 m), the army must have occupied the total road width. How long such a column must have been is difficult to estimate, but when Flavius Josephus mentions three legions (the campaign against the Jews in 67 AD) and even four (in 70), the length must have been many miles. Scouts were probably sent ahead to keep the roads free. Military traffic always had right of way over civilian traffic; when an army unit, a legion or several legions were on the road, this must have caused large traffic congestion.[546] In Late Antiquity, Ambrosius describes some soldiers who were obliged to travel along a prescribed route, because it was only there that they had the right of accommodation. After three days of travel, they had one day's rest.[547] In cities, soldiers could sometimes rest for four days or longer.[548] In such a situation, traffic must have been blocked dramatically,[549] especially in cities situated further away from the frontiers and not able to sustain huge flows of military traffic. In frontier areas, with a lot of military traffic, the *castra*, the *castella* and the cities had to have streets with enough capacity to sustain the armies and their equipment. This could explain the wide streets of Xanten.

If we assume a total amount of 500,000 military men and 100,000 km of paved roads, we get an average of five men per kilometre. Of course, soldiers were not spread throughout the whole Empire; they were concentrated in cohorts, *alae* and legions and, during the Principate, in the frontier provinces. From the third century, there were more soldiers and troop movements, but the chance of coming across an army was small.

Road security

To protect travellers against highway robbery, the government had to ensure security on the road. For this purpose, several kinds of road watchmen were active, not only within but also outside the cities. These watchmen were also responsible for the payment of taxes and catering to the requirements of the *cursus publicus*. Some groups of road watchmen will now be briefly discussed.

The *stationarii* were soldiers, placed in *stationes* (in this case 'police-posts') on crossings,[550] in cities[551] and on imperial estates.[552] Suetonius describes how the emperors Augustus and Tiberius appointed soldiers for road security against highwaymen.[553] Their task was not only to combat robbery, but also to detect runaway slaves[554] and to collect tolls and import charges.[555] At first, these men did not yet have the name of *stationarii*; this word first occurs in the first century AD.[556] We also find the word *stationarius* in Late Antiquity, in law texts[557] and epigraphical sources.[558]

Of higher level,[559] but also responsible for road security, were the *beneficiarii*. These were soldiers who were let off certain fatigue duties as a favour. They were also placed in *stationes*; we find them especially in frontier regions[560] and province capitals, where they worked for province governors.[561] One could find the *stationes* of the *beneficiarii* in the neighbourhood of *castella, stationes* of the *cursus publicus*, crossings, bridges, toll passages and cities.[562] Even if they did not have the title of *beneficiarii*, soldiers were sent to supervise places with a lot of traffic.[563] According to Nelis-Clément, these soldiers were in charge of law and order in cities, the control of taxes and the sending of messages to the authorities concerning troubles elsewhere.[564] Byzantium was strategically an important city and one can assume that Trajan considered the local security of the highest importance. There is disagreement as to whether *beneficiarii* could also be *stationarii*; this did not occur often.[565]

The most intensively guarded roads were those along the *limes*. We know that especially frontier roads in the north of the Empire were equipped with watch towers, placed at regular distances and within sight of each other. Examples of such road sections were at Hadrian's Wall in Britain, the *limes* between the Rhine and Danube in south-west Germany and – recently known – the part of the road between Utrecht and de Meern in the Netherlands.[566] Highwaymen and other offenders did not have many chances here. Nevertheless, however well the functionaries tried to maintain law and order on the roads and highways, banditry remained a problem throughout Antiquity, impossible to eliminate. Especially in periods of instability, for instance in the civil wars in the first century BC, an increase in robberies can be noted. In times of peace, it was also dangerous on the road. Pliny the Younger describes how the knight Robustus and his friend Atilius Scaurus, travelling along Via Flaminia, went missing,[567] and in 151–152, also in a peaceful period, the *legatus exercitus Africani* M. Valerius Etruscus was robbed and plundered in Africa.[568]

However well organised road security was, the *stationarii* and *beneficiarii* were also powerless against the increasing insecurity on the roads during the third century, when not only highwaymen but also troops of emperors and usurpers, invaders and other bandits – e.g. the so-called Bagaudes – were wandering through the Empire. Farms were surrounded by stone walls and staff and servants were armed; cities erected new city walls.[569] In the fourth century, there was also insecurity on the roads; travelling on your own became a dangerous pastime.[570]

Summary and conclusion

Together with the communications couriers, the army was one of the main groups for whom the road-system was realised in the first instance. The army had to move rapidly and we can assume that it indeed did so. When an army, containing some legions, was on the march, it was a long column, taking up the complete width of the road and with a length of several miles. This caused much traffic congestion, albeit for a relatively short time; the chance of meeting such an army was small and mostly a problem in the frontier regions. In Late Antiquity we see an increased army, but with a legion decreased to 1000 men. The mobility of army units thus grew, as well as the presence of army units along the roads.

The *stationarii* and *beneficiarii* were responsible for road security. They were part of the army, but their task was more police-like; to maintain law and order on the roads. Their posts (*stationes*) were mostly situated at crossings and in cities. Besides their task of fighting highway robbery, which must have been an important problem in times of civil war before the reign of Augustus, they also had to assist in the collecting of taxes.

Goods transport

Introduction

The greatest but least documented traffic is goods transport: the daily supply of goods to cities and villages. No civilisation, however primitive, can exist without the daily supply of food and materials, and Roman cities were no exception.

The majority of goods transport was carried by ship. Compared with river or sea transport, road transport had so many disadvantages that the Romans avoided it as much as possible. Cargo transport was slow and expensive. Roads were used only when there were no other alternatives. This was also the case for shepherds with their flocks in transhumance. It was unnecessary to use the broad, paved roads and highways and one could make use of *calles*, goat paths.[571]

In this section, an overview will be given of the means of transport which were available for carriers, and an attempt will be made to estimate the volume of cargo transport. Heavy transport – especially building materials – will be discussed in the next section.

Numbers of inhabitants

To assess the volume of goods transport, we must first make an estimation of the number of inhabitants of the Roman Empire. The following estimations are global, as are nearly all demographic numbers of Antiquity.[572]

The estimations of the total number of inhabitants of the Roman Empire vary from 50 to 90 million.[573] The biggest city and capital, Rome, numbered between 1 and 1.5 million inhabitants;[574] Alexandria, Antioch, Carthage, Pergamum, Ephesus and (in Late Antiquity) Constantinople had between one thousand and several hundred thousands of inhabitants. Then came the bigger province cities, also in the west part of the Empire, each with several tens of thousands of inhabitants.[575]

Which goods were transported by road?

The largest and most well-known means of (food) transport were the fleets of corn ships, sailing annually from the corn-exporting provinces of Egypt and Africa to Rome. Bulk goods like corn and olive oil were commonly transported by ship. The majority of the cities were situated on the coast or along an easily navigable river, so supply of these bulk goods was not a problem. These products also had the advantage that they could be stored for a relatively long time. However, perishables, like vegetables, meat and eggs, had to be produced in the immediate vicinity of the city to keep the interval between production and consumption as short as possible. For these products ships were also used, but mostly one depended on road transport. Road transport was considerably more expensive than sea or river transport, partly because of the maintenance of the roads.[576] Besides food, there was – of course – also need for textiles, building materials such as iron and wood and live merchandise such as cattle and slaves. Money and precious metals like gold and silver were carried by the *cursus publicus*. Products from outside the Roman Empire – from northern Europe, central Asia and central Africa – were imported at appointed places; thus Carnuntum on the Danube was the import and collecting-centre of amber.[577]

The relationship between city and countryside

No city in the pre-industrial era could exist without the countryside; the countryside was dependent on the city. Cities needed agricultural products, but they had to have a market-place where these products could be sold. Besides that, the city was the economic, social, cultural and religious centre for the countryside. The fact that cities needed many commodities and produced only a few gave rise to the well-known thesis, quoted in particular by Finley, that cities were 'parasites', and Rome was the biggest 'parasite'. The earnings of the city did not come from trade and industry, but from taxes from the whole Empire.[578]

Agriculture in Italy

The higher circles in Rome grew their own agricultural products on their estates. The common inhabitants of Rome, however, were dependent on what was sold in the market-places. Originally, corn was produced in the immediate vicinity – like all other agricultural products – but from the end of the third century BC the large-scale production of corn in the provinces commenced. Sardinia, Sicily and, in later centuries, Africa and Egypt took over the role of corn production, which caused, inevitably, a decreasing amount of corn production in Italy.[579] The enormous increase in Rome's population was, for the most part, a direct consequence of this state of affairs: many impoverished farmers had to go to Rome, trying to establish a new life. Wine production also had the same fate. Extensive vineyards in Gaul supplied cheaper wine than the small farmers in Italy. However, up to the middle of the first century AD, Italy remained an important food producer for Rome.[580]

The rivers in Italy were short and the water level was irregular, and so they were not very suitable for larger ships (the Po river was an exception). However, the drainage canal of Via Appia was used by barges. Nero also made plans for the digging of a canal from Ostia to Rome.[581] There was no alternative to road transport, sometimes combined with sea transport, for instance in Apulia.[582] This road transport could be carried out, because of the mountainous land, by pack animals, but the introduction of a paved road-system and the use of large and heavy wagons made it possible to produce more cheaply.[583]

The same situation must also have occurred in other parts of the Empire, where there were no rivers at all. Cato mentions that routes by ship, by sea and river, were important for goods transport.[584]

The production of the **suburbium** *of Rome*

The region around Rome, changing from its former production of corn to the production of other agricultural products like vegetables, fruit, flowers, dairy products and meat – that is to say fresh only for a short time – had the name *suburbium* (Fig. 2.7). Cicero mentions this name for the first time in his work *Philippicae*.[585] Although it was a densely populated area, it would be wrong to compare it with our suburbs nowadays; the little cities and villages were spread over a distance of some 10–20 miles (15–30 km) from Rome.

It is significant that all of the production centres of the *suburbium* are situated on important roads. The development of the *suburbium* could thus not take place until Rome had a properly functioning and paved road-system. The importance of a good infrastructure for the optimum functioning of a farm is indicated by Cato and again emphasised by Varro and Columella.[586]

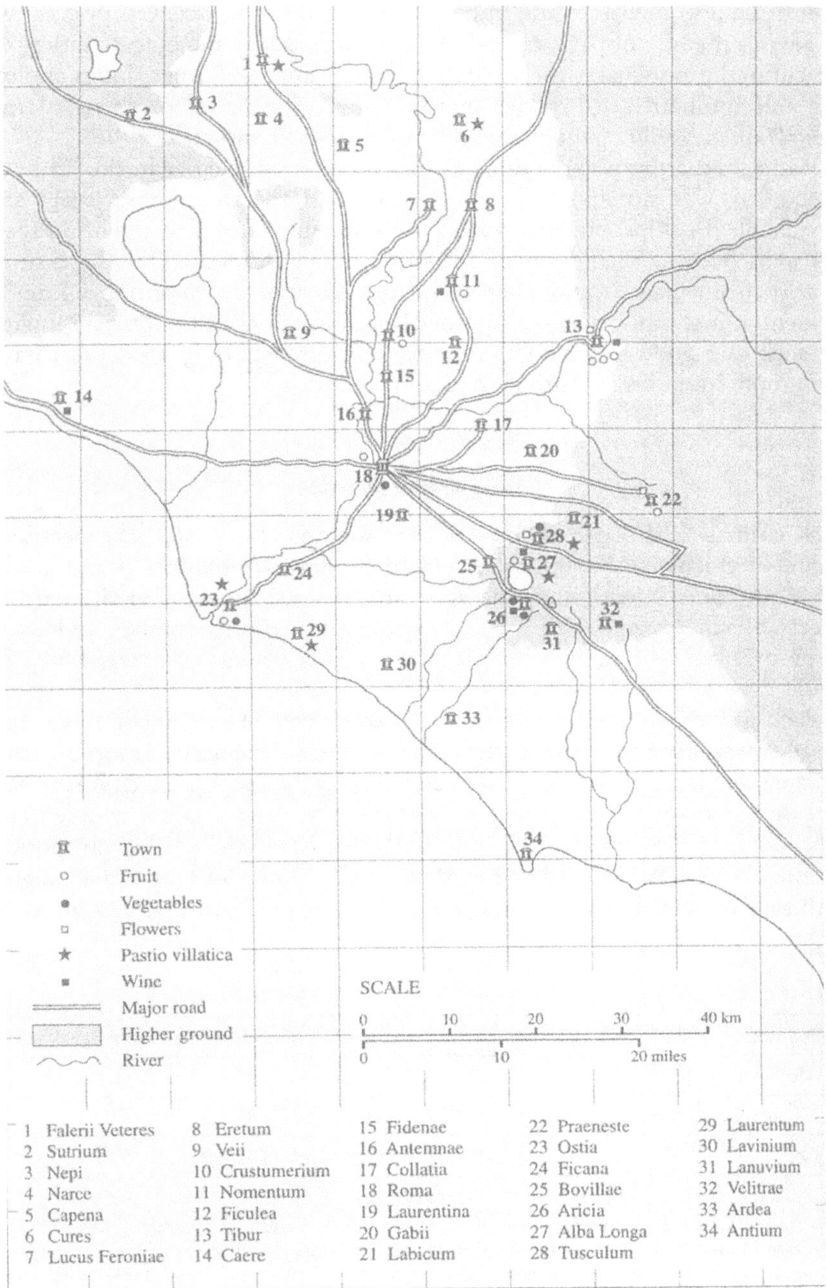

Figure 2.7 The *suburbium* of Rome. Morley 84. Cambridge: Cambridge University Press.

Columella also mentions the importance of good roads to keep prices low; he advises the hire of pack animals.[587] Tivoli produced the biggest variety of agricultural products for the whole *suburbium*, amongst other things apples, figs and mulberries.[588] Morley gives an overview of all products of the *suburbium*, as well as giving references to ancient literature.[589]

Rome's situation was the same as that of other large cities in the Empire. Alexandria was situated on the sea coast and therefore more easily approachable for ships, but Antioch was considerably more dependent on road transport. During the famine of 362–363, Julian ordered 100,000 measures of corn from Chalcis and Hierapolis, at a distance of 50 and 100 km.[590] Evidently, it was not possible to order large quantities of food by ship; with the lack of a good water infrastructure, cargo transport by road must have been more intensive in Antioch than in Alexandria.

Animals and wagons

Pack animals had some advantages over wagons. They were cheaper than wagons (which, in the end, also had to be drawn by animals) and could travel not only in mountainous agricultural areas, but also in the narrow streets of villages and cities. For instance asses, mules, hinnies and oxen come to mind.[591] The weights that these animals could carry are shown in Table 2.3.[592]

Pack animals, however, also had disadvantages. Their maximum cargo capacity was lower than that of draught animals, so one needed more animals for the same load, giving rise to higher food costs. A second disadvantage was that they were unable to carry large and/or heavy cargoes. Loads unable to be split like furniture, building material and wine barrels had to be transported by wagon. A relief from Langres (Gaul) (Fig. 2.8) shows a heavy wagon with four wheels, drawn by two mules, with one wine barrel of 525 litres.[593]

Table 2.3 Animals and maximum loads

Ass with panniers	150–200 pounds[a] (70–90 kg)
Mule with panniers	300 pounds (136 kg)[b]
Horse with panniers	400 pounds (182 kg)
Ox[c]	400 pounds (182 kg)

[a] White is not using Roman pounds, but English (453.6 gr).

[b] It is noteworthy that a mule can carry twice as much as an ass. I presume that in the case of mule and horse the weight of the rider is also included. Also strange is the maximum load weight of a horse, compared with a *cursus publicus* horse (100 Roman pounds, or about 30 kg); see p. 60, s.v. 'pack-horse'. These numbers roughly correspond with those of Raepsaet, quoting 100 to 200 kg per animal (1996–, 1098).

[c] According to Erdkamp, the functioning of an ox as pack animal was restricted; I accede to this statement. For discussion of this matter see Erdkamp 70–1, and in particular 88.

Figure 2.8 Wagon with wine barrel, relief from Langres. White 133. London: Thames & Hudson.

We also find pictures of wagons with barrels on the columns of Trajan and Marcus Aurelius; in the flat Danube regions these barrels were preferable to amphorae.[594] Draught animals could pull considerably more as well as heavier loads; a couple of horses could pull a ton (1000 kg). For this purpose, however, asses, mules and oxen were more commonly used.[595]

A point of discussion is the question as to whether four-wheeled Roman wagons had a revolving front axle. According to White, there is absolutely no concrete evidence. However, there is evidence that other people, e.g. the Persians and the Celts, had four-wheeled wagons with revolving front axles.[596] Moreover, wagons with solid front axles were unable or hardly able to drive in ruts. In any case, wagons with heavy loads must have been able to drive round bends in cities, because certain indivisible loads could only be transported in heavy wagons.[597]

Even in the case of passenger transport, many wagon names have Gallic/Celtic origins, as well as our word for 'car' (*carrus*). In Gaul, there was far more experience with wagons than in Italy. The reason was probably the landscape. Gaul, a mostly flat area, was suitable for wheeled wagons; regions like Sicily and Greece were more mountainous and more suitable for pack animals.[598] After the conquest of Gaul, during the early Principate, Gallic wagon types were introduced and improved in Italy.[599]

The most commonly used wagon for cargo transport was the *plaustrum* or *plostrum*. There is disagreement on the Gallic origin of this word.[600] The *plaustrum* usually had two wheels (solid wheels) and was drawn by a couple of draught animals; oxen were able to pull the heaviest *plaustra*.[601] A *plaustrum maius* was a cargo wagon with four wheels.[602] In some cases, *plaustra* were also suitable for passengers. The *Lex Julia Municipalis* speaks about the use of *plostra* – apart from goods transport – by the Vestal Virgins and priests and in cases of triumphal processions;[603] these *plostra* are probably not wagons for cargo purposes, but large decorated wagons. In the army the *carrus* was used, a cargo wagon with two or four wheels, drawn by four mules.[604] In his article 'Vehiculum' Lafaye mentions further the following wagon types for goods transport: *benna, carpentum, carrago, carrus, esseda, petorritum, plaustrum, rheda* and *sarracum*.[605]

A rough estimate of the logistics

It is impossible to give an exact estimate of the amount of goods traffic movement supplying a city. A rough estimate is also speculative; we simply do not know how many and by how many vehicles goods were transported. Larger wagons caused less transport movement. However, a cautious attempt to estimate the supply will be given.

Junckelmann states that a legion consumed 6 tons of corn and 5 tons of barley daily.[606] A legion numbered some 6000 men; the daily ration of a soldier can be estimated as 1 kg of corn. A city of 20,000 inhabitants should then consume, according to this information, 20,000 kg (20 tons) of corn. But in a city we are talking of a population including women, children, etc. So the daily consumption of a city should be estimated at 10 tons. When we assume that this city is not situated on the coast or along a navigable river and all corn had to be imported via road by pack animals, the daily transport of corn could be divided into seventy-seven daily transports per pack animal, with an average load of 130 kg. If porters (*saccarii*) were used instead of pack animals, or if the maximum weight for the animals was not reached, the number of logistical movements was, of course, higher.

By what factor the total amount of traffic movements carrying other products must be multiplied is impossible to say. Animal food, building materials, raw materials (e.g. clay), textiles, food, wine, etc. could, however, be transported by wagon and in many cases also by pack animals, but which vehicle was most in use depended on many factors. Wagons were confronted with traffic-restricting measures like the *Lex Julia Municipalis*, whereas pack animals were otherwise free of it. In cities situated in mountainous regions wagons would not have played an important role. In cities there was, of course, not only traffic movements relating to imports, but also export of goods and waste, where wagons and animals were also used; an empty wine barrel used just as much space as a full one. This also explains the presence of deep wheel ruts in the neighbourhood of inns in Pompeii.[607] The amount of goods transport movements must have run into hundreds, at its busiest along the roads between the harbour and the industrial quarter, where in many cases the shops were also situated. In a city like Pompeii, where the majority of the streets were narrow, goods transport by wagon was subjected to one-way traffic that was forced to follow fixed routes, hampered by stepping stones, barricades and – when supply had to take place by night – poor street-lighting or none at all. That traffic between harbour and city, in both directions, must have involved pack animals and *saccarii* in northern Europe also is apparent from the fact that the harbour gates of Xanten were too narrow for large wagons. In chapter 4 the traffic policy of Xanten will be discussed.

Corn distribution in Rome

Probably the most crowded place in Rome was the vicinity of Porta Trigemina, where river ships unloaded for the warehouses and granaries (*horrea*).[608] From there distribution to the shops had to take place, probably by means of pack animals and *saccarii*.

Beneficiarii, people who were entitled to partake in corn distributions (*frumentationes*) – the first corn distributions took place in 58 BC[609] – received a monthly quantity of 5 *modii*, about 35 kg.[610] Two texts indicate that this distribution took place on the same day each month. The first is a passage from the *Legatio ad Gaium* of Philo of Alexandria, where it is stated that Jewish *beneficiarii* could receive their monthly ration the following day when the day of the *frumentatio* coincided with the Sabbath.[611] The second is a passage from the *Lex Julia Municipalis*, where the announcement of the *frumentationes* is mentioned.[612] The number of *beneficiarii* in Caesar's time was 320,000, when one assumes that the *Lex Julia Municipalis* was also introduced; he reduced it to 150,000.[613]

Both the aforementioned texts, however, are not conclusive evidence that the *frumentationes* actually always took place on the same day of the month. Such a system could not have functioned very well; on one single day, a very large amount of corn (375 tons)[614] from the granaries on the Tiber side had to be transported to several points by means of transport which was not used on all other days of the month. In Samos, the corn distribution took ten days; travellers who came back later could then collect their ration.[615] Rome, with its high population density and poor infrastructure, needed even more extended logistics. According to Virlouvet, the Circus Flaminius, the Saepta Julia and the Porticus Minucia Frumentaria were the distribution points.[616] Further, the *Forma Urbis* mentions a *statio annonae* at Forum Boarium.[617]

At the end of the third century wine, oil and pork as well as corn were distributed to the *beneficiarii*.[618] This measure caused even more traffic and, combined with the Aurelian Wall, more traffic congestion around the gates.[619]

Summary and conclusion

Goods transport accounted for enormous traffic flow, which is poorly documented. In literature, the transport of agricultural products is mentioned only incidentally by authors such as Cato, Varro and Virgil. This transport took place mainly by ship. Long distances could be covered, especially by sea, so goods which kept longer, like corn, could be transported by ship. Perishable products like vegetables and eggs had to be produced in the immediate vicinity of the cities; the better the roads, the longer the distance from which such products could be carried and the larger the food-producing hinterland.

Goods transport by road mainly involved the use of pack animals, especially in mountainous regions like Italy and Greece. For small farmers these animals had many advantages; they were relatively cheap and could move everywhere. Wide roads were not easy to build and it was unnecessary to build them merely for agricultural transport; for smaller pack animals, narrow goat paths (*calles*) were sufficient.

Wheeled vehicles had more advantages in flat regions; the origins of the majority of wagon types are Gallic. Because wagons were expensive to buy and to maintain – draught animals were also necessary – they were appealing to larger farms and estates as they could transport heavy loads. Cities in flat regions with a good road-system (especially in allocated land), like Gaul and northern Italy, could easily be supplied in this way. In cities with a sufficient infrastructure, there must have been room enough for large and heavy wagons and few traffic restrictions. In the event of restrictions, however, these cities also required pack animals and/or *saccarii*.

Special transport

Introduction

In this section we will consider three special forms of traffic, which have not yet been discussed. These forms of traffic were incidental, but our knowledge of them is relatively good. They are:

1. the journeys of the emperor and his court, becoming more important in the third and fourth century especially;
2. the transport of animals and predators for the games;
3. heavy transport: the transport (under supervision) of building materials which could not be transported by wagon, for instance marble.

The emperor and his court

The journeys of the emperor throughout the Empire – on campaigns or for other purposes – can be seen as a combination of travel by the army and the *cursus publicus*. A long column was on the road, the cities had to supply lodgings for many people and they even had to subsidise them, and other traffic, including food supply, was disturbed.[620]

The emperor's visit to a province was such an important event that this was announced months beforehand; *mansiones, stationes* and cities could then provide for enough stocks. Alexander Severus therefore announced his travel route several months beforehand during his campaign against the Persians.[621] Tiberius also prepared his voyages by announcing traffic procedures; wagons were collected and traffic through the cities was regulated.[622] Emperors on campaign, for example the aforementioned Alexander Severus, had a large army, which needed sufficient equipment.[623] Caracalla, however, announced voyages which he never made.[624] The number of quartermasters who had to prepare the emperor's voyages was, according to Giebel, about 5000 men and took a year to accomplish.[625]

The functionary responsible for the organisation of the emperor's voyages in Italy was the *praefectus vehiculorum* and outside Italy was the province governor. Subsidiary to him were the *aedilis castrensium*, responsible for the food supply for the emperor and his court, and the *acceptor vehiculorum*, responsible for the collection of wagons. The function of *aedilis castrensium* was introduced by Augustus and probably existed – like that of *praefectus vehiculorum* – only in Italy. Under Domitian, it was probably Plotius Grypus, the man to whom Statius dedicated his ninth poem of the *Silvae*, who was his *aedilis castrensium*.[626] Under Marcus Aurelius, there was a *praepositus copiarum*, Ti. Claudius Candidus, responsible for food supply. Besides the function of *praepositus copiarum* there was also the *praepositus annonae*, with the same significance as *praepositus copiarum*.[627] Later, in the third and fourth centuries, these functions disappeared and the management of the voyage, food and accommodation supply and the management of the logistics would have been the responsibility of the *cursus publicus*.[628]

Every emperor preferred his own means of transport. Augustus was happy to use a sedan, preceded by an *explorator viae*,[629] fulfilling here in fact the function of a lictor and comparable with the whip-bearers, of whom a mosaic has been found in Ostia (discussed in chapter 3). The soldier-emperor Caracalla preferred marching on foot, together with his troops.[630] The wagon was the most suitable vehicle when the emperor was accompanied by his family (*consessus vehiculi*). As punishment for a defeat, Galerius was forced to walk alongside or next to the *carpentum* of Diocletian.[631] Other emperors travelled on horseback.

There were also various lodgings for the emperors, from a building constructed from wood and loam for Julian in Batnai, via sanctuaries to a villa for Nero in Olympia.[632] The more luxurious lodgings were usually located, from the second century onwards, in cities situated at strategic points, for instance Nicomedia and Sirmium. The emperor's lodging outside the city was called a *palatium*.

During his voyage, the emperor was surrounded by many people. To this group belonged, first, his family and friends, imitating the Hellenistic kings (φίλοι and ἑταῖροι) and the Republican *cohors amicorum* of the governors. Augustus founded the *consilium principis*; its task was to advise the emperor.

To this council belonged friends, senators and knights. During the voyage, the chancellors received the title of *comes Augusti*. Besides the *comites Augusti*, there were also subordinates such as writers, artists, heralds and doctors. The most famous doctor of the Roman Empire, Galen, accompanied Marcus Aurelius and Lucius Verus during their voyage to Aquileia.[633]

How many did the emperor have in his company? We know that Caesar, on his visit to Cicero's villa in Campania, had an entourage of 2000 men consisting of soldiers, slaves and freemen.[634] Hadrian was accompanied by about 5000 on his way to Egypt; this number is estimated from the amount of food which was necessary: amongst other things 3000 bundles of hay, 372 pigs and 2000 sheep.[635]

Extravagant emperors also mobilised many wagons and citizens for their travels. Caligula required so many wagons from Rome for his voyage through Gaul that there was no more bread for sale (because wagons were unsuitable in Rome by reason of the lack of wide streets and the restrictions of the *Lex Julia Municipalis*, one must assume that these wagons had to distribute whole loaves); Nero's entourage contained more than a thousand wagons and (according to Giebel) a hundred she-asses for their milk, needed for Poppaea.[636] Still more extravagant was Heliogabalus; he would have used more than 600 wagons, 10,000 camels (!) and 500 *carrucae*.[637]

It is clear that a travelling emperor caused enormous organisation, comparable with an army on the march, and from the end of the second century AD onwards an emperor was often on campaign. Not only the traffic, but also the whole economy was disrupted. One had to find lodgings, vehicles and food for men and animals. The consequences were, in many cases, famine. The famine of Antioch, which has already been mentioned, caused by Julian's troops in 362–363, is notorious. There was not only a lack of food, but also enormous water consumption in the city.[638] Only with the abundance of water in Antioch and help from the emperor himself could total chaos be avoided.

Julian imitated earlier emperors. Athens received corn from Hadrian and Tarsus from Caracalla and Alexander Severus when their armies passed the city.[639] Another emperor who according to literature tried to save the local population was Trajan; in his *Panegyricus*, Pliny the Younger devotes much attention to this.[640]

Wild animals in transport

Another category deserving the name of special transport was that of predators and other large animals, used mostly for the arena games. These animals transported themselves or were transported in cages. Besides games, unusual and exotic animals, like elephants, were also used in triumphal processions.[641]

Elephants. The Romans knew two categories of elephant: the African and the Asiatic. Remarkable here is that the Asiatic elephants were bigger than their African counterparts.[642] These African elephants were a – now extinct – subspecies of the African elephant. Their habitat was the so-called Maghreb, north of the Sahara (now Morocco, Algeria and Tunisia); they were also to be found in the Horn of Africa.[643]

Although some nations used them in battles – for instance Pyrrhus, Hannibal and the Indian king Porus – the Romans did not use them for this purpose on a large scale; Claudius however made use of them on his campaign in Britain, during the conquest of Colchester.[644] In the first century AD, there was a herd of elephants in Laurentum, the emperor's private collection.[645] They were probably kept as a hobby, for use as a gift to friends and relations. The Persians also used elephants in battle, but not successfully. Gordian III had thirty-two elephants in Rome, twelve of which were supplied by himself. Before his assassination in 244, the Senate had already given him permission for a triumphal procession, and on this occasion his triumphal car was drawn by elephants in honour of his victory over the Persians.[646]

The transport of elephants could cause problems. Elephants, however, could move around on their own, but they could not go everywhere, certainly not in cities and settlements with narrow streets and gates. This was a problem in Rome. Pompey wanted to enter Rome with his triumphal car drawn by four elephants, but according to Plutarch the gate was too narrow.[647] Because he uses the word πύλαι (plural), one assumes that a gate with more than one passage is meant, maybe Porta Trigemina; this gate was situated directly next to the Tiber bank and elephants could easily be unloaded when they came by ship from Ostia to Rome. Heliogabalus seems to have successfully reached Rome with his quadruple of elephants, but during his procession tombs in the Ager Vaticanus were damaged; evidently the road was too narrow.[648] Aurelian climbed Capitol Hill during his triumphal procession with twenty elephants.[649] Nevertheless, a quadruple of elephants must have been possible; a medallion of Tiberius (Fig. 2.9) shows a wagon drawn by four elephants, carrying a wax statue of Augustus.

Figure 2.9 Medallion with a quadruple of elephants. Reinach 542. Paris: Hachette.

More evidence comes in the third century, when the Senate agreed to Gordian III having a triumphal procession with a quadruple of elephants; evidently, at that time (244 AD), there was enough space for such a quadruple. However, quadruples required wide roads and these were scarce in Rome's inner city: only Via Sacra, Via Nova and, possibly, Via Triumphalis were options. It is certainly possible that during the Empire the approach roads were too narrow for quadruples and that the elephants entered the city in single file; once in the city, they could be harnessed up where enough space was available. The number of streets where elephants could move was, however, very limited and substantive evidence of elephants entering the city as a quadruple is unfortunately difficult to find.

Hippopotamuses. Not much is known about hippopotamuses. In Antiquity, they came from the western part of North Africa, like the aforementioned elephants,[650] Egypt and Palestine. Hunting these animals was considered difficult and dangerous. The first mention of a hippopotamus in Rome is in 58 BC, at the games of M. Aemilius Scaurus.[651] Heliogabalus and Gordian III had some hippopotamuses in their private zoo.[652]

Rhinoceroses. In Antiquity, there were two known categories of rhinoceroses: the Indian and the African. The Indian one was named by Pliny the Elder 'Indian ass'.[653] The African rhinoceros was shown, according to Pliny, at the games of Pompey, in 55 BC.[654] A lot of pleasure was to be had from seeing a rhinoceros fighting an elephant; in the third century AD, this was a recurring arena game.[655] This is also evidence that relatively many elephants were imported and used in this way. The rhinoceros was well known in the first century AD. Martial compared men with rhinoceroses. Gordian III had one rhinoceros in his stable.[656]

Giraffes. During the triumphal procession of Caesar in 46 BC, Rome made acquaintance with the giraffe.[657] Perhaps he was imitating Ptolemaeus II, who made use of παρδάλεις for a triumphal procession;[658] it is not clear, however, whether these animals were panthers or giraffes. Horace gives a description of an animal in Caesar's triumphal procession, half panther, half camel; one assumes that it was a giraffe.[659] Giraffes are also mentioned in the *Historia Augusta*; in the tally of exotic animals in the zoo of Gordian III (see above) there were also ten giraffes. They were given away, like the aforementioned elephants, hippopotamuses and the rhinoceros, or slaughtered during the games.[660]

There were special cage wagons for the transport of more regular-sized animals: deer, bears, boars and possibly also lions, tigers and panthers. These cage wagons were not suitable for the bigger and heavier animals. In Figure 2.10 a cage wagon is shown; the mosaic was found in Piazza Armerina, Sicily. It also shows that such a cage wagon is no larger than a *plaustrum*. Another similarity with the *plaustrum* is that the wagon has solid wheels. The vehicle was drawn by oxen. Wild animals ordered by the imperial court

Figure 2.10 Cage wagon. Pace Fig. 2.1. Roma: Gherardo Casini.

were transported by the *cursus publicus*, the *cursus clabularis*. This was also the case in Late Antiquity.[661] In the cities, the aediles were responsible for the transport of animals for the games.[662]

Was it possible, especially for the bigger animals, to be transported everywhere? It has already been mentioned that Pompey's elephants could not pass a city gate in Rome. A modern, adult Indian elephant has a length of 5.5 m and a width of 1–1.5 m.[663] The average road (*via publica*) had a minimum width of 8 feet (about 2.40 m) and would not have caused problems. But in Rome, the majority of streets were narrower than 8 feet; they did not have the status of *via*, apart from some exceptions like Via Nova.[664] Big animals like elephants could only walk along special routes in the city. Quadruples of elephants, like that of Pompey, could rarely pass a city gate. When we consider the dimensions of the Indian elephant, we arrive at 6 m for a quadruple. Of all the researched city gates in the following chapter, only the central passage of Porta Praetoria in Aosta is wider than 6 m.[665]

One can assume that the biggest animals could also move or be transported through other cities without many problems. In the Romanised western part of the Empire, especially where more amphitheatres were built and more arena games took place, more animals were transported. In Rome, however, not only did more arena games take place than in other places in the Romanised west, but they were also on a larger scale; there were also triumphal processions requiring the transport of big animals to be regulated along prescribed routes, the majority of the streets in the city being too small for elephants or rhinoceroses.

Heavy transport

Finally, so-called heavy transport. The definition of 'heavy transport' used here is the following: indivisible loads of more than 1 ton, unable to be transported by wagon. The maximum load capacity of an *angaria* was 1500 pounds, about 500 kg. In this section, we will restrict ourselves to discussing

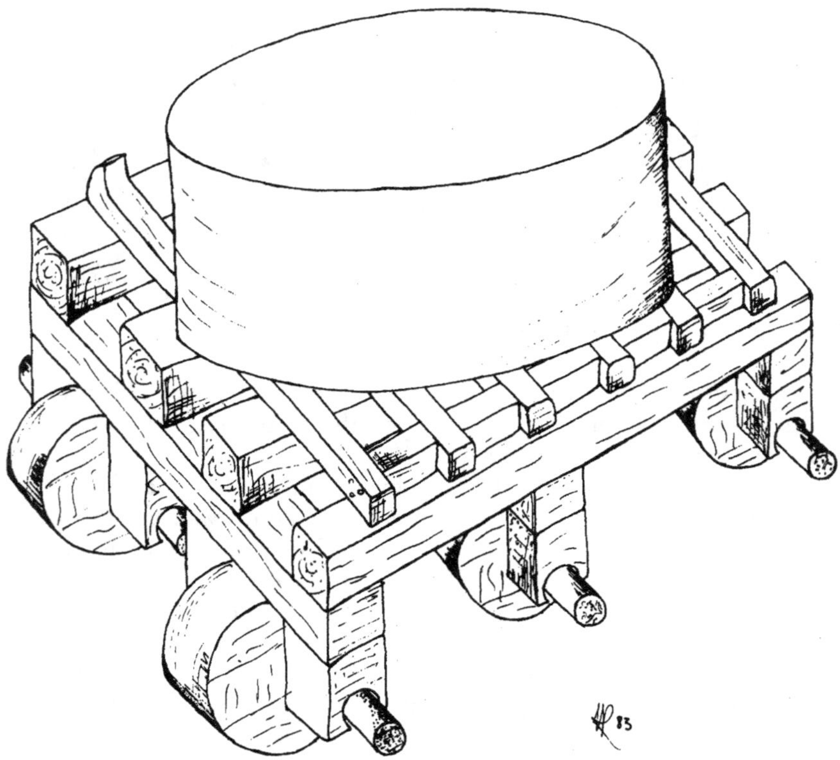

Figure 2.11 Reconstruction of a vehicle for transport of column parts. Raepsaet 1984, 125. Louvain: Istas.

the transport of indivisible loads, too heavy for any wagon. Heavy transport also encompasses trees and tree-trunks (where possible, they were transported over water), heavy stone blocks and column parts.

The Greeks already used special vehicles for the transport of column sections. Raepsaet shows in his article on the transport of column sections (from Pentelicon to Eleusis) a reconstruction illustration of a very heavy vehicle with six wheels (Fig. 2.11); every wheel has its own axle and double axle suspension. In an inscription from 330 BC[666] a vehicle is mentioned with seventeen axles and sixty-eight pins; such a vehicle would have been able to carry a load of 7.5 tons.[667] Another Greek vehicle is the so-called *ἅμαξα τετράκυκλος*.[668]

The aforementioned special vehicles for heavy transport were Greek, but we must assume that the Romans also used these vehicles. Apart from that, they developed special new types. In Egypt, a vehicle with twelve wheels was used. These were probably made up of six pairs of wheels, each with its own axle.[669] A vehicle with twelve (spoke)wheels, with a thickness of about 50 cm should, technically, be able to carry a load of more than 200 tons.[670]

The heaviest loads transported by Romans were the monolithic obelisks from Egypt. The heaviest was an obelisk from Karnak, with a weight of 500 tons. The obelisk was brought to Rome by Constantius, almost completely over water. In Rome, the obelisk followed the route through Porta Ostiensis, via Vicus Piscinae Publicae to the Circus Maximus.[671]

As draught animals oxen were, of course, most suitable, but asses and camels also were used. For the transport of loads of 10–15 tons of marble, one needed twelve to eighteen pairs of oxen; this means of transport was in use until the eighteenth century.[672] DeLaine states that the speed of this heavy transport was 1.67 km/hour.[673] However, one granite column (200 tons) needed 400 to 450 asses.[674] To avoid a shortage of oxen in the agricultural sector, heavy transport took place by preference in summer in Greece and winter in Egypt;[675] here one had to bear in mind the rising of the water in the Nile.[676]

Conclusion of chapter 2

Generally, one can state that passenger traffic in Antiquity, particularly long-distance traffic, did not take place on a large scale. There was passenger traffic *intra* and in some cases *extra muros*. This traffic was, together with the transport of goods over a short distance for the supply of the city, the most prolific. However, it is also the least documented. We learn nothing, or next to nothing, from literary sources; all our information is based on archaeological sources and research on statistical, logistical and economic information. Traffic by land was avoided as much as possible when transport by ship was available, in spite of the good quality of the paved road-system. This was, of course, mostly the case with heavy transport; an obelisk from Egypt travelled thousands of miles over water (Nile–Mediterranean–Tiber) and only a few miles over land.

We are better informed about specific traffic over a long distance. The army and the *cursus publicus*, the two special groups for which the paved road-system was built and to which it owes its existence, made up a much smaller portion of the total amount of road-users. A third group, comparable with these, is created by the emperor and his court. One could say that the quantity of information we have about road-users decreases as the volume increases. The *cursus publicus*, the Roman state transport company, must have accounted for a relatively minor part of the total logistics. Its famous (or infamous) reputation was, however, caused by other factors, namely its importance for the state and the severe charges which had to be paid by the population.

The volume of traffic was not set. In Late Antiquity, the volume and type of traffic differed from that during the Principate. We see a change in importance, from the cities at the centre of the Empire to cities on the periphery, and a decrease both in the size of existing cities and in the number of inhabitants. In many regions, especially in the West, trade traffic decreased in the interaction between city and countryside; instead, estates became more self-supporting. Another reason for this development was the decreasing safety on the road.

State traffic also increased, caused by the army, the *cursus publicus* and the emperor himself. Instead of remaining in solid *castra* and *castella*, the army became more mobile and the cavalry more important. The increasing bureaucracy also caused an increased *cursus publicus* which, along with transport function, also became a security medium. It is doubtful if the two types of traffic, army and state transport company, would have increased the traffic problems. But they caused problems for the local population, who had to supply wagons, animals, food and money.

One who travelled on his own usually walked or rode. The most popular animals to ride were the ass, mule and hinny, especially in mountainous areas. Horses were scarcer and oxen more suitable for goods transport. In deserts, dromedaries were used. Coaches and wagons were also in use; anyone who happened to want to travel could hire a vehicle near a city gate. One disadvantage of private possession of a vehicle was that not all axles in the Empire had the same width. Wheeled vehicles were actually only suitable in flat areas and were too expensive for the majority of people; moreover, draught animals were still necessary.

We must assume that for most transport the advantages of vehicles were outweighed by their disadvantages. Only for large, heavy and indivisible loads was a vehicle strictly necessary.

3

TRAFFIC CONGESTION

In chapter 1 it was stated that the Roman road-system was of a high quality and in chapter 2 that the volume of the traffic was relatively small, certainly too small to cause much traffic congestion on long-distance roads – with some exceptions. Does this mean that traffic congestion was an unknown phenomenon in the Roman Empire?

Delay could have resulted from other reasons than imperfections in the road-system, as we have today. Road repairs, accidents, breakdowns and highwaymen were regular problems for travellers and could be considered as a *force majeure*. There was only one category of delay for long-distance travellers caused by the government: passing toll points.

Besides the long-distance delay caused by toll points, a bigger problem was caused by the local delay on the edges of and within the cities; notably the organically developed cities in the Mediterranean area (particularly Italy) could not sustain traffic without problems and drastic measures had to be taken to keep them accessible. Infamous bottlenecks were city gates: many gates had only one single passage, which traffic entering and leaving had to pass through. Other gates, usually built in the Late Republic and Principate, had more passages and so were more suitable for traffic. This is mostly visible in the newly founded cities, the so-called *coloniae*, with wide streets and a chessboard street plan, comparable with a *castra*.

After passing the gate, the traveller came in amongst the city crowd, which one had to wrestle through. The city with the worst traffic congestion was without doubt Rome, where the unfortunate combination of its extreme population density, a chaotic and insufficient street system and its metropolitan nature combined. However, this situation was unique.

In this chapter, an overview will be given of traffic congestion in the Roman Empire; this took place for the most part on the edges of and within the cities. Attempts to solve these problems will be discussed in chapter 4.

Toll points

Introduction

The Latin word for 'toll' is *portorium*, a word connected with *portus*, 'harbour'. The toll by far the most levied was a form of harbour tax. Here too the idea of harbour toll is adapted from Greece, where it was known as ἐμπορίαι.

Toll was levied at several points. Besides harbour tolls, there were also tolls along roads, at river crossings, province boundaries, mountain passes and cities. The most expensive tolls had to be paid at the frontiers of the Empire. There were tolls for the maintenance of urban facilities and there were tolls where the revenue was paid to province and state. It depended on the place and the emperor whether and how many tolls were introduced or dropped.

In literature, there are many references to tolls and toll points. By far the most are related to financial and organisational aspects of tolls; again, we know considerably less about the logistical aspects. To obtain this information, we again need archaeological sources.

The situation of toll points

Tolls levied in harbours and on the frontiers of the Empire are not the subject of this book and will not be discussed here.

The first citation of *portorium* is by Livy, who writes that the population was exempt from it. This exemption took place in 508 BC, two years after the abandoning of the monarchy.[677] This makes reference to the fact that the population had to pay *portorium* and *tributum* before that time.

There was a toll at the city gates of Rome at the end of the second century BC, and in other cities in Italy conquered by the Romans.[678] In 60 BC all tolls in Italy were abandoned according to the *Lex Caecilia*, but to finance their wars, Caesar and other *triumviri* not only reinstalled them, but also increased their quantity.[679]

In the time of the Republic, there were certainly *portoria* in the provinces. In his *Pro Fonteio*, Cicero mentions four toll points for wine transport in Gallia Narbonensis: in Toulouse, Crodunum, Vulchalo and Elesiodunum. One had to pay *portorium* at every point along the route.[680] According to de Laet, these toll points, called *stationes*, were situated along Via Domitia; one must assume that these were placed along the road. Moreover, here there was an exception: Cicero called this 'tax policy' a *crimen invidiosissimum*, a 'scandalous crime'.[681] Nevertheless, as time passed more and more toll points were set up. At the time of the transition from Republic to Empire, there were *stationes* in cities, on rivers and at crossings.[682]

During the reign of Tiberius tax reforms took place, whereby districts were combined. Commodus increased the number of toll points; his successor Pertinax subsequently abandoned these in 193. As well as in the harbours, these new points were placed along roads and riverbanks.[683] Further, there

Figure 3.1 Bridge with portcullis. Masquelez and Saglio 967. Paris: Hachette.

were tolls at province boundaries (in contradiction to the situation in the Republic) and at the boundaries of tax districts into which the Empire was divided.[684] Here the officials preferred natural boundaries like river crossings and passes across or through which traffic was forced to travel. We know that at least one bridge was equipped with a gate with portcullis (Fig. 3.1).[685]

An important place for the levying of tolls was Lyons. The presence of a boundary between the provinces of Gallia Narbonensis and Gallia Lugdunensis, and the confluence of the rivers Rhône and Saône, made this city very suitable for the installation of toll points. There was toll levying at city gates, the evidence being the finding of pieces of lead that were attached to tax-free goods.[686] In the Alps, there were also many tax-collectors;[687] their presence here can be explained by the high cost of road building. Further, there were many toll points in the Danube provinces (see Fig. 3.2).

In Rome tolls were levied in certain cases. To earn revenue, Caligula created a tax on food, lawsuits and prostitution. Officers of the Praetorian Guard were ordered to collect these taxes.[688] Pliny the Elder mentions a *portorium* on vegetables,[689] and in the times of Marcus Aurelius and Alexander Severus there was levying of a further tax, *ansarium*, as shown in some inscriptions; this tax was collected on the approach roads to the city.[690]

Palmer states that the gates in the Aurelian Wall were erected mainly in places where taxes on goods had already previously been collected. Tolls are mentioned along Via Flaminia, Via Salaria, Via Asinaria, Via Appia, Via (Campana)-Portuensis, at Porta Esquilina and a hitherto unknown point; after Aurelianus outside Porta Praenestina, Labicana and Tiburtina and, in the fifth century, at Porta Nomentana, which replaced Porta Collina.[691]

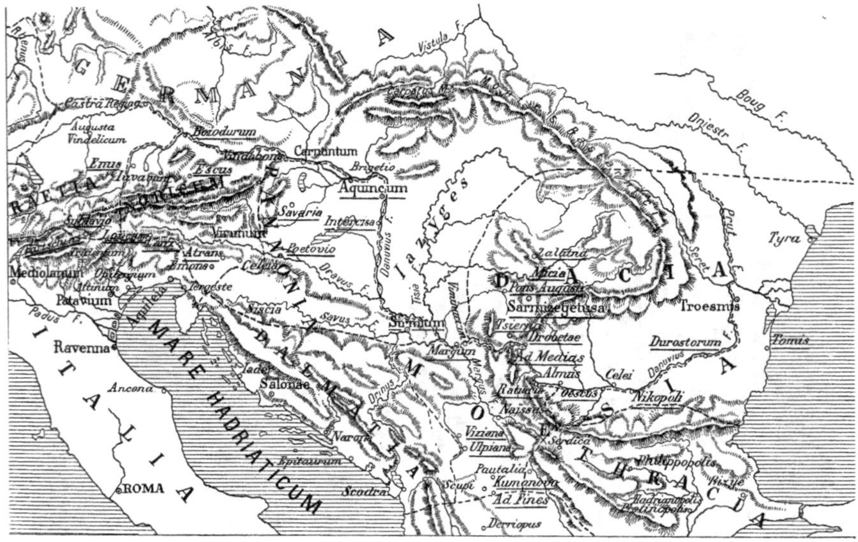

Figure 3.2 Toll places in the Danube provinces. Cagnat 589. Paris: Hachette.

Aurelian erected the wall not only as a defence structure, but also to prevent smuggling.[692] To distinguish them from the *portae* in the wall, the tax collecting points were called *cancelli*.[693] Because the devaluation of money was gigantic and the *beneficiarii* received not only corn but also wine, tax had to be paid not in money but in wine. The 'tax in wine', *fiscalia vina*, was collected in wine cellars on the west bank of the Tiber.[694] These wine cellars were, in fact, the only buildings outside the Aurelian Wall without a funeral function.

Licences of tax-collectors

Certain groups and goods were exempt from toll. Cagnat lists the following groups:[695] the emperor and his court,[696] diplomats,[697] officials of amphitheatre games,[698] ship-owners,[699] province governors, other high state officials and, in Late Antiquity, palace functionaries.[700] Soldiers, veterans and their families were also exempt, certainly during the reigns of Nero, Domitian, Constantine, Valentinian I, Valens, Gratian and Justinian.[701]

Not only certain persons, but also certain goods, were toll free. Whether goods for own use were also exempt is a point of discussion. Quintilianus mentions that toll (one fortieth) had to be paid over all goods, except the vehicle itself and everything belonging to it (*instrumenta itineris*).[702] On the other hand, a passage of the *Codex* of Justinianus states that agricultural instruments and goods for own use (in contrast to trade goods) were tax free.[703] This passage seems to be more credible, amongst other things by the finding of pieces of lead in Lyons.[704] Further, all goods for the army and the taxes themselves were exempt.[705] All other goods were not tax free; toll was even levied on the transport of dead bodies.[706]

The aforementioned citation by Quintilianus shows that tax-collectors (*publicani* or, in Greek-speaking regions, *τελῶναι*) had licence to inspect passing goods, and any which were not declared immediately were confiscated. To exercise these licences, the inspections of the luggage of passing travellers (this was done by assistants of the *publicani*, the so-called *scrutatores*)[707] must have caused considerable delays, especially at busy points. A traveller like Aelius Aristides, with his long entourage of slaves and luggage, must have suffered under these inspections. Perhaps at certain places (maybe everywhere), the *publicani* got help from soldiers or guards at the toll point. There was a toll point in Ehl, south of Strasbourg, where a detachment of the Legio VIII Augusta was present for the guarding of the local *statio*.[708]

A problem for the *publicani* must have been that it was not always clear which items were for private use and which for trade. The tax-collectors were forced to investigate all people, animals and goods meticulously, impeded by the law to overlook *matronae*.[709] Plutarch furiously renounced the indiscrete behaviour of the tax-collectors; he called them 'busybodies' (*πολυπράγμονες*).[710]

Probably the most unpopular tax officials were the praetorians of Caligula, levying money for all food sold in Rome. Suetonius is not clear exactly where the *centuriones* and *tribuni praetoriani* were positioned to collect money but in the crowded city, this must have caused long queues of people waiting. On the other hand, those who had to pay tax on animals (*foricularium*) could park them in a pen or corral beside the tax point. A herd of sheep, goats or pigs could cause unnecessary blockage.[711]

Summary and conclusion

The majority of toll points were situated on the frontiers of the Roman Empire and in the harbours. At these points, most economic activities took place. Besides that, there were also toll points along roads (especially at bridges and passes) and at gates of cities; open cities, like Rome was before the building of the Aurelian Wall, could have points along the road where tolls or other taxes could be levied.

Legal passages and the finding of pieces of lead indicate that some groups and goods were exempt from toll or other taxes. The most expensive toll had to be paid by people travelling on their own and traders. When we consider the long distances between the majority of toll points, the chance of passing more than one toll point in one day was not great.

In passing a toll point with many people and/or a lot of luggage one had to allow long delays. The *publicani* and *τελῶναι* had the right and even the duty to investigate meticulously all road-users and their luggage, with the help of their assistants.

City gates

Introduction

In this section attention will be given to the city gates. City walls and gates were the most essential structures of a city. They guaranteed the safety of the inhabitants, at times when a city could be under attack. Until the nineteenth century, cities everywhere in the world were surrounded by walls and canals, with the gates as the accepted entry-point.

Roman cities were no exception. Because some of these cities are very well preserved, we have a good knowledge of the methods of gate building and why they were built so. Again we are more or less completely dependent on archaeology, because there is hardly any attention to gate building in literature. Even our most important literary source for architecture, Vitruvius, does not give attention to the way gates had to be constructed. Fortunately, archaeologists have found – as well as gates which have survived the tests of time and sometimes maintain their original appearance – many traces of former gates. So we can deduce that there is a clear evolution in the development of gate building.

To what extent did gates cause traffic congestion? Gates were always the biggest source of congestion; sometimes the traffic could be blocked completely. Because a gate was always a part of a defence structure and so could be closed, one can also conclude that gates were specially built to more or less stop the traffic, when the situation required it. To that end, the gates had first to be constructed in a certain way; there had to be certain equipment in place to close the gates effectively by means of doors, portcullises or a combination of both. Besides that, there had to be men available to work with this equipment.

In the first part of this section, the architectonic aspects of gates will be discussed, where the evolution from narrow to large gate complexes and returning to narrow gates again will have a prominent role. The functioning and opening of the gates, and the gate-keepers, will be discussed in the second part.

The situation and building of city gates

Planning of city gates

The building of the defence structures, including the city gates, was the basis of the city itself. The boundary of the city was in fact the boundary of the defence structure.

The Romans' manner of boundary planning is derived from the Etruscans. A description of this is given by Varro: a furrow was made in the land by a plough. This happened on a day which was approved by the gods.[712] At the points where roads crossed the furrow, the plough was lifted; here the gates had to be built.[713]

The line situated inside the wall was called *pomerium* (*post murum*, *postmoerium*);[714] the strip of land between this line and the wall also got this name. Later the strip of land outside the wall was also called *pomerium*.[715] Inside the *pomerium* the *genius* (city god) of the new city was in force. In chapter 4 the *pomerium* will be discussed in more detail.

The number of city gates depended on the number of approach roads and not on the number of inhabitants or the size of the city boundary. Seen from a military perspective, it was desirable to keep the number of gates as low as possible; the gates were the weak points of the city defence. Narrow gates, easy to close, were preferable. These interests are in contrast to economic interests; for traffic, many and large gates were preferable. During the course of the history of the Roman Empire not only gates with military importance were built, but also those with economic importance, which were more comfortable for traffic.

As stated, little is written on gate building. Vitruvius concentrates on city defence in only one chapter and here only strategic aspects are mentioned. In one section, the approach roads are discussed, likewise in a strategic–military context.[716]

The first gates of Rome

In earliest times, Rome was a small settlement on the Palatine Hill. We know that this settlement, known as Roma Quadrata[717] had a defensive wall of 1500 m[718] and three or four gates.[719] Varro gives the names of three of them: Porta Mugonia, Porta Romana and Porta Janualis; the last one was the famous gate consecrated to Janus, always open in times of war, but closed in times of peace. A fourth gate is mentioned by Giovagnoli, consecrated to the Magna Mater Cybele.[720] Etruscan cities usually had three gates, a holy number; they were consecrated to the three main gods Jupiter, Juno and Minerva.[721] In Volterra, a gate is still visible with three heads, presumably depicting these three gods (Fig. 3.3).[722]

Livy describes how the sixth king of Rome, Servius Tullius, built new fortifications around the city, so the *pomerium* also had to be extended.[723] These fortifications, generally known as the 'Wall of Servius Tullius' or 'Servian Wall', date, however, from the fourth century BC and were erected by the Gauls after the sack of Rome, which took place in or around 390 BC. Giovagnoli states that the Servian Wall was erected between 378 and 352 BC.[724] The total boundary of the fortifications totalled some 12 km: 1350 m of it as an earthen wall (*agger*) and the other parts as stone walls.[725]

Figure 3.3 Volterra, Porta all'Arco. Scullard 1967, 68. London: Thames & Hudson.

How many gates these fortifications included is not exactly known. According to Stambaugh, there were thirteen gates,[726] Palmer mentions eighteen (just as the *Forma Urbis*),[727] Giovagnoli twenty[728] and Pliny the Elder no fewer than thirty-seven.[729] Säflund also discusses this problem. The total number of gates she mentions is twenty-five, including '*XII portae*' in the vicinity of the Circus Maximus.[730] So, one arrives at thirty-seven gates, he same as Pliny's number. However, it is doubtful that twelve gates were concentrated in a part of the wall no longer than the short flat side of the Circus Maximus, the departure point of the charioteers. One possibility is the following translation of Pliny's passage '*portas, quae sunt hodie numero XXXVII ita ut Duodecim semel numerentur*': 'The number of gates is 37, where Twelve-Gates is counted as only one' and not 'The number of gates is 37, where Twelve Gates are counted as only one', the translation of Rackham (Loeb edition);[731] moreover, *Duodecim* is written here with a capital letter.[732] It could be very possible that here was one gate in the city wall, in the vicinity of the twelve gates of the Circus Maximus.[733] Moreover, one must realise that the Servian Wall dates from the fourth century BC, a time when Rome had continuous conflicts with other tribes and cities in Italy; the number of gates to defend had of course to be kept as small as possible. This situation changed when after the Third Punic War the threat of war with other cities was decreased and Rome had to sustain more traffic. In Pliny's time, long sections of the Servian Wall had already disappeared, and were covered up by new buildings[734] (so the wall's course intersected, amongst other things, the Forum Julium, even the Temple of Venus Genetrix);[735] he mentions that seven gates had disappeared, not including the thirty-seven above-mentioned gates.[736]

Not only is the number of gates not known exactly, we also do not know much about their appearance. They do not seem to have been very large; not only was there not much traffic yet, but the city defences were still important. Säflund shows a plan of Porta Viminalis, placed in the *agger*-section of the wall. The passage was a little more than 3 m wide. Apart from that, this gate was not important.[737]

If we follow Rackham's edition of Pliny and its implication that there are twelve 'double gates',[738] these gates could have been constructed like Etruscan ones and as are also found in Pompeii. These gates are constructed as two vaulted sections – an inner gate and an outer gate at each side of the wall – with an open passage or courtyard between them.[739] Closing took place by means of one or more doors. An example of this type of gate is Porta di Stabia in Pompeii.[740] 'Double gates' must here be understood not as gates with two passages next to each other, but as one behind the other.

At any rate, Rome had one gate with more than one passage. As regards the name of Porta Trigemina, situated at the Forum Boarium, we must state that this gate had three openings. Perhaps there was a gate complex with three openings, two on one side of the complex and one on the other and in between them a courtyard. According to Lyngby, Porta Trigemina has three passages next to each other; a reconstruction illustration, with three passages of equal height, shows a gate in the Hellenistic style. This view is strongly doubted by Brands, who bases his misgivings on the scarce archaeological sources.[741] The fact that this gate was a larger one can be concluded by the fact that here there was very busy traffic.[742]

Another gate, probably a double gate, was Porta Carmentalis, situated south-west of Capitol Hill. Dionysius of Halicarnassus mentions that there was talk of 'gates', plural;[743] according to Livy, there were two passages, *dextro iano portae Carmentalis*.[744] But as with the aforementioned problem of the '*XII portae*', we have a linguistic problem, in this case the words *dextro iano*. One usually translates 'they departed through the right gate of Porta Carmentalis'. Holland suggests the translation 'they departed through the passage, right of Porta Carmentalis'.[745] Säflund states that the Porta Carmentalis was not a double gate in the fourth century BC[746] and assumes that the gate survived as a monument, after the disappearance of the city wall.[747] So this gate was, at any rate, not an obstacle for the traffic in this area; Porta Carmentalis was not far away from Porta Trigemina. That is the reason why it is not impossible that this gate could have been a double gate.

The gates of Pompeii

The oldest phase of the Samnitic city defence dates from the fourth century BC, the same time as the Servian Wall. Later, this defence was continuously upgraded,[748] as well as the seven city gates: Porta di Ercolano, Porta del Vesuvio, Porta di Nola, Porta di Sarno, Porta di Nocera, Porta di Stabia[749] and Porta Marina. Excavations have finally given definite evidence that Porta di Capua did not exist.[750] Apart from Porta di Ercolano and Porta Marina, they are all narrow gates, not able to sustain busy traffic, but kept narrow for a good defence.[751] Only one wagon was able to pass at a time, but the connecting streets to the gates were wider.

Porta Marina has a parallel-running side-gate for pedestrians, built immediately next to the main passage. Porta di Ercolano is constructed in a totally different way. The gate is larger and has a side-gate for pedestrians on each side of the central passage for wagons and horsemen. It is a gate which could cope with busy traffic, where the traffic flow was divided by type. This sort of gate (in the following text called a 'three-passage-gate') dates from a later time,[752] when there was more emphasis on a good flow of traffic. Another difference is that the central passage of this gate could be closed by a portcullis (Fig. 3.4).[753]

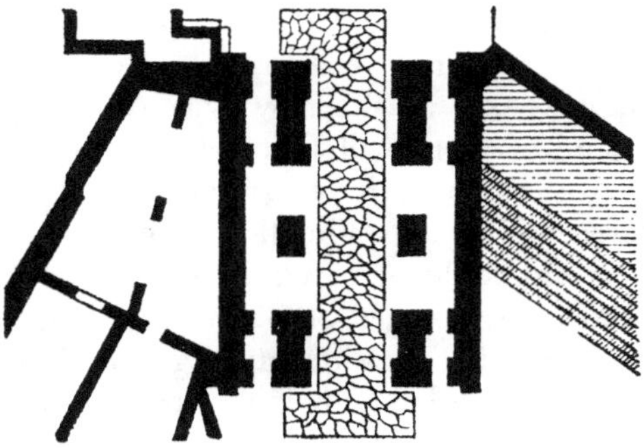

Figure 3.4 Pompeii, Porta di Ercolano. Mau 237. New York: MacMillan.

Traffic accommodation at the gates of Rome and Pompeii

The above text shows that the passages of the gates of Rome and (most) gates of Pompeii were so narrow that only one wagon could pass at a time. At first, this should not have been a problem. In the fourth century BC, streets had not yet been paved. The first paved street in Rome was the Clivus Publicius (238 BC).[754]

Later, when traffic increased, more paved roads were built and the gate passages were also paved. To give traffic the opportunity to wait for oncoming vehicles the decision was made to pave forecourts, not only on the city side, but also on the land side. Porta Viminalis had such a forecourt on the city side.[755] In Pompeii, these forecourts (in fact street broadenings) are visible, amongst other places, at Porta di Stabia, Porta del Vesuvio and Porta di Nola. Porta di Ercolano was broadened on both sides, in spite of its very wide passages: the main passage in the centre has a width of 4.4 m and the pedestrian gates 1.35 m.[756] The forerunner of Porta di Ercolano was situated more to the west, probably a single gate like the other gates of Pompeii.[757]

New cities, new gates

From the first century BC onwards began a period when many cities were founded. Because these cities – most of them acquired the status of *colonia*[758] – were planned according to the system of Hippodamus, like a *castra*, with a chessboard pattern and no need for large-scale defence structures, this allowed for a change in emphasis to socio-economic interests. The *coloniae* were founded for several purposes, including the housing of veterans and other colonists and the Romanising of the surrounding region.

A good infrastructure was essential now and the accessibility of these cities had to be as good as possible. Instead of massive earthen and stone walls, occupying much space, we now see smaller stone walls as defence structures and boundaries. The *coloniae* were founded in conquered areas, so the building of large-scale fortifications was not necessary.

As well as new defence walls, we see also a new type of city gate; more detailed descriptions of some of them will follow below. They are imposing gate complexes with two, three or four passages next to each other, especially built for different types of traffic flow. These passages are sometimes completely vaulted and could be closed by portcullises; this required the building of a floor above the passages. Some gate buildings have a courtyard. Another new element, typical of this category of gate building, is towers flanking the passage; they sometimes have a height of tens of metres.

The origin of this new style of gate is found, according to Schultze, in the Hellenistic kingdoms and especially in Alexandria.[759] Hellenistic cities were mostly open cities; only a few had stone walls as defence.[760] Unfortunately, we know little of Hellenistic walls and gates. Of the Alexandrian gates, to which Schultze refers, we only know the names, Sun Gate and Moon Gate,[761] and the probable situation of the two main gates. The archaeologist Empereur, however, states that these gates, situated at the far ends of the main east–west connection, were built under Hadrian or Antoninus Pius.[762] But we can find a good example of a gate in situ, dating from the fourth century BC, in Assos (Asia Minor). This single gate, flanked by towers, also shows how a passage was built without knowledge of the arch construction; the upper side is triangular and looks like the Lion's Gate of Mycene.[763] Another gate of this type, still in a good condition, is visible in Messene (Peloponnesus).

Not only were these new gates more functional because of the larger amount of passages, they could also function as political propaganda for the local authorities. Gates, visible from far away, could be distinguished and impressive buildings and were in many cases elaborately decorated.

The number of gates of a *colonia* was restricted. The main approach roads, the *cardo* (running north–south) and *decumanus* (running east–west), intersecting each other at the centre of the new city, had each its gate in the surrounding wall. Thus the number of gates was restricted to three or four. Because *cardo* and *decumanus* were also the places where most economic activities *extra muros* (suburbs, troublesome workshops) took place, it was especially here that gates with three or four passages were constructed.[764] Here specifically was busy (pedestrian) traffic.

Besides these gates, cities could also have more gates for in- and outgoing traffic. The roads through these gates were less important and the gates were themselves generally, simpler.[765]

Everywhere in the Roman Empire, especially in the west part of it, traces, remain of gates and even almost entire complete gates are found. The most famous example is the Porta Nigra of Trier, but in Autun also one can still see remains of four-passage-gates. The north-west gate of Xanten, Burginatiumtor, has been reconstructed. Here follows short descriptions of various gates.

Figure 3.5 Fréjus, city gate. Schultze 292. Mainz: von Zabern.

Fréjus

In 44 BC Fréjus acquired, from Caesar, the status of *colonia* and had four gates. The gates of the *decumanus* can be considered as forerunners of the new style of gate planning; the city wall had here a curve within the city boundary and formed a semi-circular forecourt, with two towers on the corners. In the centre of the forecourt, the *decumanus* ran to the gate complex itself (Fig. 3.5).

A gate planned in this way had practical disadvantages. The forecourt occupied a lot of space within the city boundary and traffic entering and leaving was forced to take a sharp bend. This concept was not imitated by later *coloniae*. According to Petit's map there is no evidence that the forecourt had the function of a square as at the gates of Pompeii;[766] probably, the space between road and city wall was part of the *pomerium*.

Aosta

The gates of Fréjus have two passages. In Italy, some cities chose – from the erection of Porta di Ercolano onwards – to construct gates with three passages, for example Spello, founded in the forties of the first century AD.[767] Shortly after 25 BC Aosta was founded, as the *colonia* Augusta Praetoria Salassorum. Porta Praetoria, one of the original four large gates, is one of the first new-style gates described above and an integral part of the city plan.[768]

The total width of the gate complex is no less than 41 m; 20.60 m is occupied by the gate itself – the central part – and 10.20 m by the flanking towers.

Porta Praetoria (Fig. 3.6) is, just as the aforementioned gates of Spello, a three-passage-gate; the front, on the city and the land side, consists of three

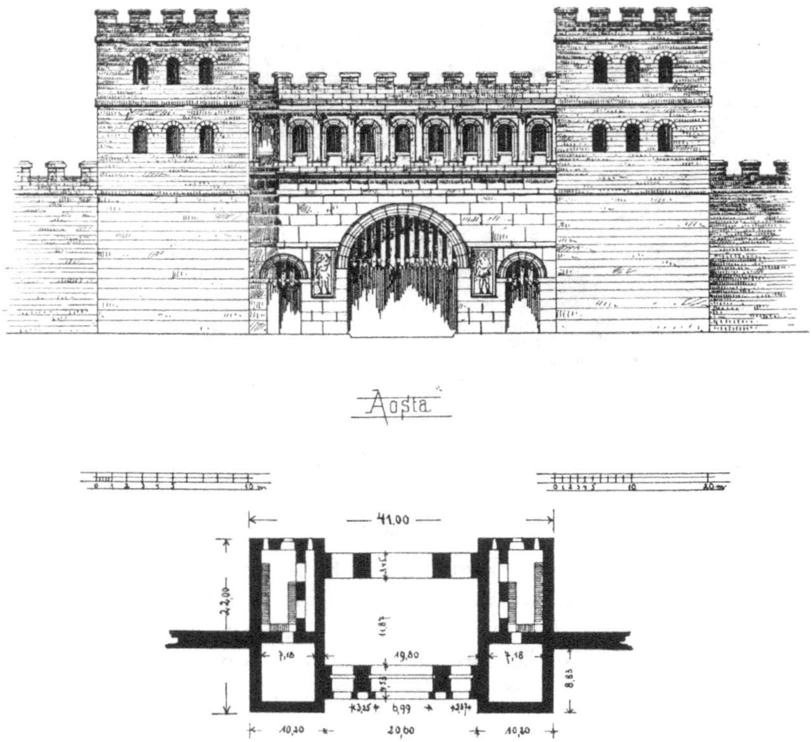

Figure 3.6 Aosta, Porta Praetoria. Schulze Taf. XIII. Mainz: von Zabern.

vaulted passages: a high, wide passage in the centre for wagons and horsemen and two smaller passages, either side of it, for pedestrians. The façades surround the rectangular courtyard together with the towers, with dimensions of 19.80 × 11.87 m. This courtyard is the largest of all Roman gate complexes at 236 m^2.[769]

It is obvious that such a construction was almost ideal. Pedestrian traffic was divided from wheeled traffic (a four-passage-gate also divided wheeled traffic entering and leaving) and gave not only safety to pedestrians, but also a minimum time to wait to pass through the gate. An invading enemy who had passed the outer gate could be kept in the courtyard by closing all doors and portcullises, which the gate-keepers could drop down from the first floor above the passages.[770] The towers also gave added protection.

Autun and Colchester

Four-passage-gates are scarce. They are found in two regions: north-west Italy and the bordering part of Gaul, and in Britain. The reason for this scarcity is clear: such gates were the most expensive to build. On the other hand, they gave the city more stature. Moreover, they were safe: all four traffic flows were divided. It is noteworthy that the Gallic-Italian gates still had a courtyard, but the British four-passage-gates had continuous walls, completely dividing the traffic in the gate-building.[771]

Autun had four four-passage-gates. Two of them are still standing: Porte St André (Porta Lingonensis) (Fig. 3.7) and Porte d'Arroux (Porta Senonica). The total gate complex of Porte St André has a width of 40.26 m and is only a few tens of centimetres narrower than the aforementioned three-passage-gate of Aosta, but here there are four passages instead of three. Another four-passage-gate, Balkerne Gate in Colchester (dating from the second century),[772] has a width of only 32 m. This shows that the width of the front is not dependent on the number of passages. But in the case of equal front widths, a larger number of passages results in thinner pillars between them, and this creates lower passages. Balkerne Gate has wide central

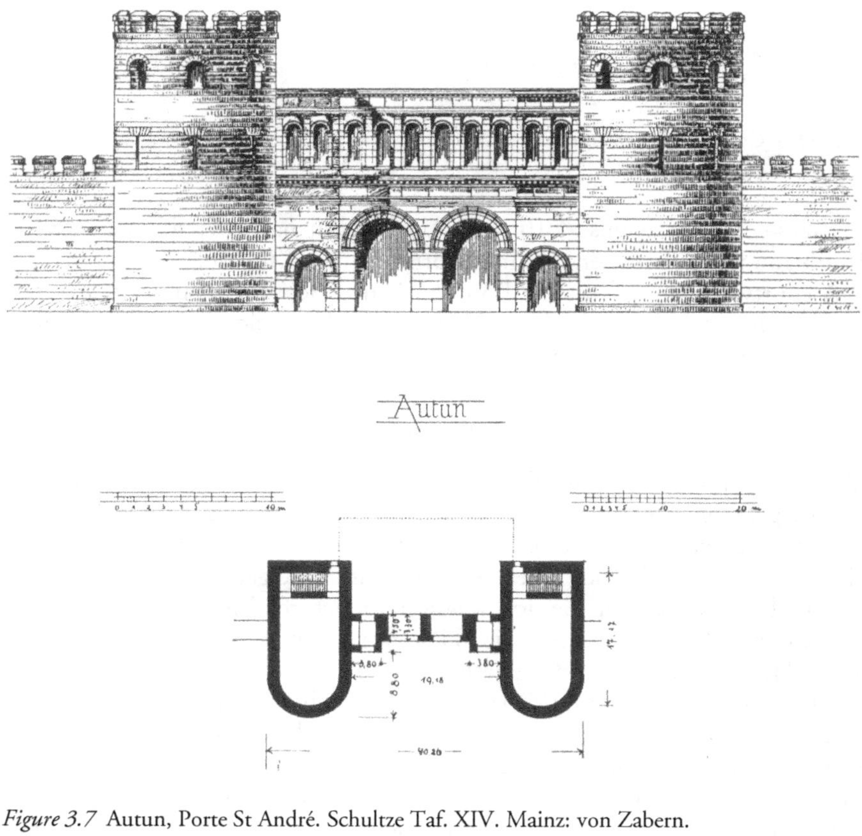

Figure 3.7 Autun, Porte St André. Schultze Taf. XIV. Mainz: von Zabern.

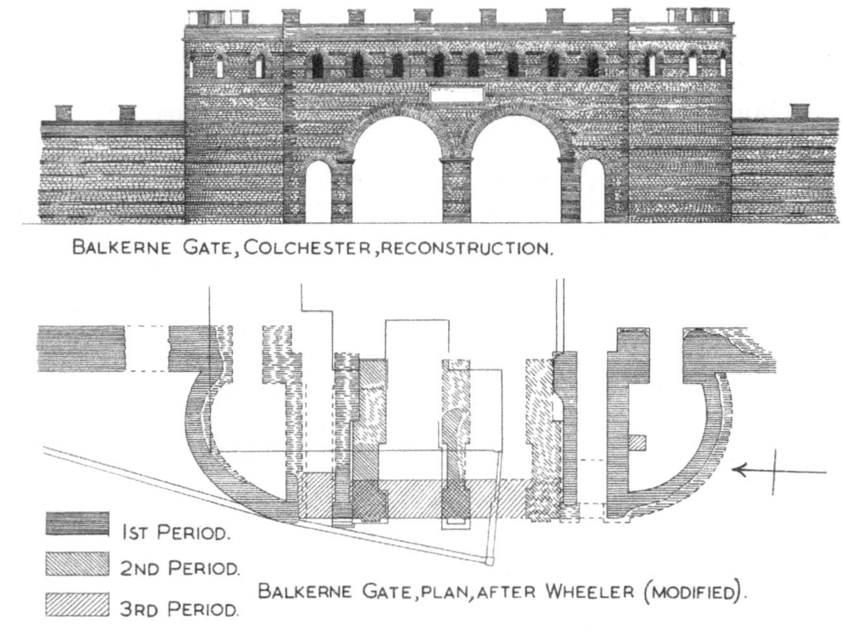

Figure 3.8 Colchester, Balkerne Gate. Hull 18. London: Society of Antiquaries/Oxford University Press.

passages – 5.10 m against 4.10 m in Autun – but narrower side passages of 1.80 m and also narrower, quadrant-formed towers (Fig. 3.8).[773]

A courtyard is not found in Porte St André. Schultze assumes that there was a courtyard, but in the first centuries AD the choice was perhaps made to build a gate without a courtyard.[774]

Xanten

Xanten obtained the status of *colonia* in *c.* 100 AD. The city would acquire a land gate on the three land sides: they were built as double gates or three-passage-gates.[775] The four harbour gates in the north-east city wall, giving access to the harbour area, were built as single gates.[776]

The double gate over the north-west approach road is completely reconstructed in the original situation. This gate is known as Burginatiumtor, derived from the *castellum* Burginatium to the north-west. With a total width of a little more than 26 m, this is one of the smaller gates found in *coloniae.* The gate does not have a courtyard, only a central pillar dividing the traffic flow, comparable with Balkerne Gate in Colchester. The first floor, from where the portcullises were operated, is attainable by the city wall and the towers, each with one door on the city side. So it is clear that the towers

Figure 3.9 Xanten, Burginatiumtor. Photo: C.R. van Tilburg.

do not have a traffic function at all; nowhere was there a direct connection between the passages and the towers with their defence and status symbol function.[777]

During the rebuilding, not only were the two portcullises reconstructed, but also the doors on the land side (Fig. 3.9).

Trier

Porta Nigra ('Black Gate', a medieval name derived from the black stones; in Antiquity the gate was known as Porta Martis)[778] in Trier is, like Burginatiumtor, a typical example of a double gate. The fact that the main gates of Trier are double gates was probably due to the fact that Trier did not have suburbs causing intensive traffic; the walls were erected at a distance from the built-up city area.[779] The gate looks heavy and cumbersome because of the short central section and the two floors above the passages. In combination with the high towers (30 m),[780] it was and remains a remarkable building (Fig. 3.10).[781]

Because the gate is very well preserved, by researching the ruts of the portcullises one could establish that the underside of them was covered with iron and impossible to lift higher than the beginning of the arches.[782]

There was, for a long time, discussion about its date of inception. Besides mythological 'dates' in Greek times, Lehner dated the building to the beginning of the second half of the third century (the reign of the usurper Postumus, 259–268). Other dates refer to the fourth century, but the most generally accepted date nowadays is the last quarter of the second century,

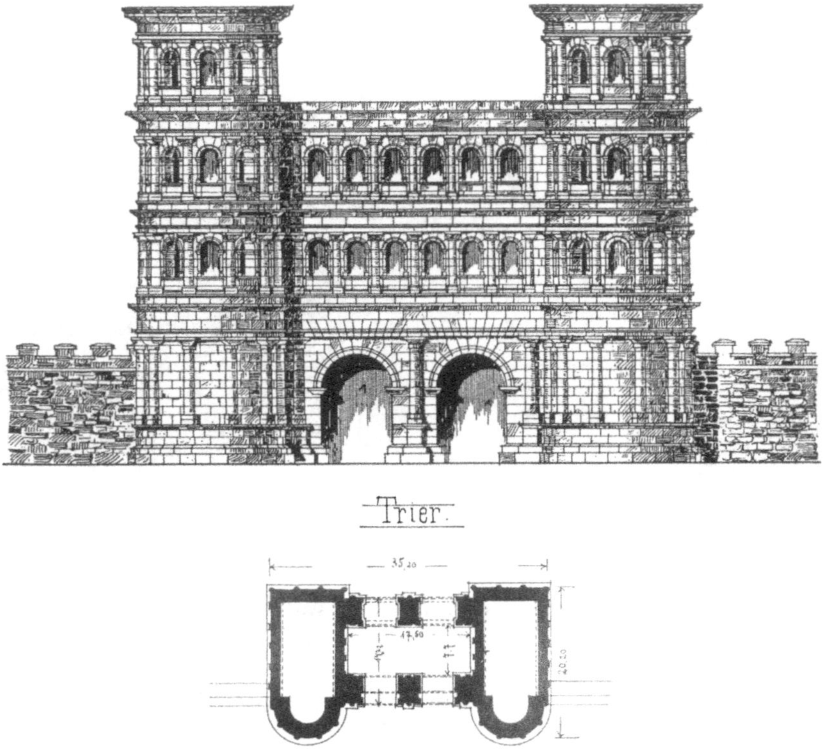

Figure 3.10 Trier, Porta Nigra. Schultze Taf. XVI. Mainz: von Zabern.

the same as the city wall. Confirmation is given by the fact that part of the wall was built in a cemetery, in use from the first century to the third quarter of the second. Apart from that, according to Cüppers, there is historical evidence: the gate was not yet complete when Trier was attacked and besieged by the Germans during the conflict between Septimius Severus and Clodius Albinus – in the years 196–197.[783]

Gates in the east provinces

Roman gates are to be found in the Hellenistic provinces, especially Asia Minor. These are quite different from their west European counterparts: there are no gate buildings, only spans. These spans are not part of a defence structure. Traces of doors and portcullises are lacking, which raises the question whether these gates could be closed. They must merely be considered as a kind of division between city and countryside; the function is more that of a triumphal arch than a city gate.

Figure 3.11 Antalya, Hadrian Gate. Schultze 329. Mainz: von Zabern.

Domitian Gate in Hierapolis, so-called because it was built in the time of Domitian (81–96), is such a gate. The gate consists of three adjacent arches, flanked by semi-circular towers. Immediately noticeable is that the passages on the left and right are almost as high as the central passage.[784] We find this type of construction in other places in Asia Minor, for example in Patara.[785] In Antalya, there was a very richly adorned gate, dedicated to Hadrian, with three passages of equal height (Fig. 3.11).[786]

In Cilicia, south of the Cilician Gate (a mountain pass) in the neighbourhood of the modern village of Sağlıklı, an arch was built over the road in the fifth–sixth centuries AD. Along the road between Antioch and Aleppo another such arch is found, again dating from Late Antiquity. The function of this gate is unclear, but Hellenkemper and Hild assume that these gates had the same function as the aforementioned gates in Patara and Antalya, but here they were erected by private individuals.[787]

Outside Asia Minor, a very beautiful and well-preserved 'city triumphal gate' is found in Timgad (Algeria); this three-passage-gate has a high central passage and narrower side passages, so it resembles the three-passage-gates in western Europe.

In Palestine, nearly all city gates date from the pre-Roman period. They were heavy, fortified single gates, part of large building complexes.[788] As well as that, there were also smaller gates in the city walls. According to an unproved statement the expression 'eye of the needle' in the New Testament refers to these small gates, only for pedestrians.[789]

Some general remarks

The aforementioned gate complexes are mostly to be found in major cities with the status of *colonia*. Because the *coloniae* were planned cities with a restricted number of gates, it was possible to finance these expensive gate complexes. The number of *coloniae* was restricted; many more cities did not have this status (for instance most cities in Gaul).[790] These cities were usually not surrounded by walls[791] and in this case they did not have gates at all. Cities with walls had in most cases simple gates, but sometimes more of them. Tongeren, in the north of Gaul, had five gates, one of which was a double gate; the other gates were single.[792] They were not large; a wide passage requires a large-scale gate building.

It has already been mentioned that four-passage-gates are to be found in north-west Italy, south-east Gaul and Britain. The double- and three-passage-gates have a wider spread area. Three- and four-passage-gates were built where economic activities took place outside the walls (or where such activities were anticipated) and where there was busy (pedestrian) traffic.

The roads did not only run through cities, but also through *castella* and *castra*. These military settlements had single and double gates which all traffic had to pass through.[793]

How was the traffic flow through these gates? In the case of three-passage-gates the pedestrian traffic, using the smaller side gates, did not have to pay any regard to oncoming traffic and so the through-flow was ideal. In the case of four-passage-gates this was also the case for riding traffic. At double gates all traffic – pedestrians, horsemen and wagons – had to reckon with oncoming traffic. In the case of deeper gates especially, with long tunnel-passages, this situation could cause traffic congestion and delay, assuming that in the case of double gates both passages were accessible to traffic and separated by direction; but there is a possibility that pedestrian and driving traffic each had its own passage, in both directions.

Late Antiquity

The period in which imposing gate complexes were built ended in the second half of the third century. The well-functioning economy of cities had gone, caused by detoriating political and economic circumstances, combined with invasions. In the frontier regions new, smaller cities came into being, sometimes inside the former boundaries (for example Tongeren)[794] and sometimes alongside (for example Nijmegen).[795] In the central regions, walls were erected around former open cities (for example Rome). In all these cases, new walls were necessary and, of course, new gates. These new gates were built initially for defence, in contrast with the aforementioned gate complexes. Outside the walls, deep ditches were dug out, interrupted by causeways before the gates; the *pomerium* was more distinctive than in former times. The gates themselves had heavy, semi-circular towers, flanking narrower passages. Of course, the importance of (wagon) transport had to make way for other, military interests, but in Late Antiquity we see a new traffic flow: the tides of mobile armies, imperial courts, etc.

During the course of time (fourth and fifth centuries), we see a steadily decreasing use of defence structures and gates, because they no longer had a function. Gates were destroyed, they fell into ruins and they were bricked up (for example Balkerne Gate in Colchester, after which the road to London was diverted to the single south-east gate or Head Gate)[796] or were given another function (for example Porta Nigra).[797]

Gates in the Aurelian Wall

Everywhere in the Empire cities built smaller defence structures, but 600 years after the Servian Wall Rome erected its longest city wall in its history. In 272, Aurelian started the building of the wall and seven years later, in 279, the work was finished by Probus. The wall had a length of 18 to 19 km (with 2 km on the right hand bank of the Tiber) and fourteen gates. The maintenance must have been enormous. In 307, Maxentius restored the wall and in 402, Honorius rebuilt the wall extensively.[798]

Most of the gates were simple single gates; they did not have a courtyard[799] and were flanked by heavy towers. Some of the gates had a forecourt.[800] Others, like Porta Asinaria, were less important and did not have towers. Four gates were built as double gates: Porta Ostiensis, Porta Appia,[801] Porta Flaminia and Porta Portuensis;[802] Porta Praenestina, in combination with Porta Labicana, was considered as a double gate, but in fact these were two separate gates, giving access to two separate roads.[803] Moreover, the wall also contained some smaller gates, named *posterulae*, of lesser importance and probably only meant for pedestrians.[804]

Honorius fortified the walls and gates in 402–403. The four double gates were still reconstructed as single gates; all gates had a courtyard (Fig. 3.12).[805] The gate immediately south of the Castra Praetoria was even bricked up completely (Porta Clausa).[806]

A typical example of such a gate, dating from Aurelian's time, rebuilt by Maxentius and fortified by Honorius, is Porta Ostiensis, nowadays known as Porta S. Paolo. We see a heavy, high gate building with roughly the same frontal width (34 m) as Porta Nigra in Trier. Aurelian built a double gate; the gate was fortified by Maxentius (he made the gate complex deeper) and Honorius rebuilt the two passages into one. The towers completely dominate the complex and the central part – including the relatively narrow passage 4 m wide in Honorius' time – is no wider than 13 m. There is no decoration at all; one can rightly say a single, plain gate, just wide enough to give access to traffic.[807]

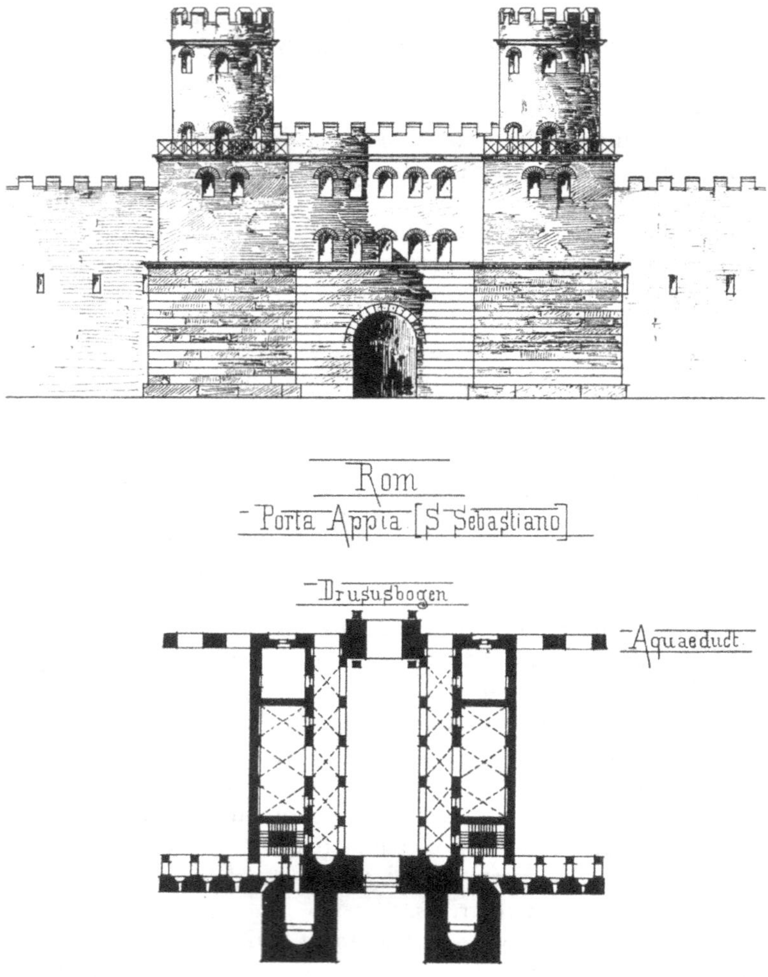

Figure 3.12 Rome, Porta Appia. Schultze Taf. XVII. Mainz: von Zabern.

Summary and conclusion

City gates represent a conflict of interests between the military and economic-logistic activities of a city. Seen from a military point of view, the ideal situation was a city with the fewest and narrowest gates possible. Road traffic, however, required a city as approachable as possible, with the most and widest gates possible.

The periods of rise, flourish, decline and fall, typical of civilisations including the Roman Empire, can be recognised in the gate constructions.

Most city gates in Pompeii still give the impression of a city in a rising period: narrow gates which could easily be closed, situated in a defensive wall offering security. Porta di Ercolano already shows the construction of a new style of gate, built not initially for defence but for traffic.

The gate complexes in the large-scale planned *coloniae* like Porta Praetoria in Aosta symbolise a period of flourishing, roughly coinciding with the Principate. These were large, imposing passages for traffic, able to be closed by portcullises and doors, which could obstruct traffic in peacetime but not an aggressive invasion by barbarians and usurpers in times of war. The gates of this period did not cause traffic congestion; traffic had right of way.

In some cities, particularly in the Hellenistic regions, we find gates with no military function; they were unable to be closed owing to their construction and had the function of a status symbol. Any closing had to be done by manpower. The function of a gate-keeper will be discussed in the following section.

In Late Antiquity, we see a return to narrow gates, easily defended, where traffic interests again had to give way to military interests. But even the heaviest walls, the smallest gates and the deepest ditches could not prevent the decline and fall and, finally, the depopulation of the cities in the west part of the Roman Empire, which is a characteristic of the Dark Ages.

The functioning of city gates

Introduction

How did the city gates really function? When were they opened and closed? What role did the gate-keepers play and did they cause traffic congestion? This matter will be discussed in this section.

Opening and closing times of city gates

We do not have much information about opening- and closing-times of city gates. We have to take note of the fact that not all cities had walls and gates. So in Gaul, cities with a surrounding wall were more the exception than the rule, especially the Celtic cities and settlements. Roman *coloniae* usually had walls, but not all of them; Trier for example did not have a wall before the end of the second century AD.[808] So walls were not the essential feature of a city, like in the Middle Ages.

So the question is: what was the function of walls exactly? Certainly, they could give protection against enemies in periods of instability, but this was not the case with all walls. The walls of Xanten were erected *c.* 100 AD, when the city acquired its status of *colonia*. At this time, there was no turmoil or war, so it was not necessary to construct heavy and thick walls as in Late

Table 3.1 Gate dimensions

City	Name	Type	Dimensions gate building (without towers)	Width of passages	Total width	Depth of passages	Dimensions of towers	Date
Antalya		three-passage-gate	width 17.75 m depth 3.25 m	all three 4.15 m	17.75 m	35.15 m	width 8.70 m depth 9.50 m	Hadrian
Aosta	Porta Praetoria	three-passage-gate	width 20.60 m depth about 20 m	central passage 6.99 m side passages 2.37 m	41 m	±20 m	width 10.20 m depth 22 m	Augustus
Autun	Porte St André	four-passage-gate	width 19.18 m no traces of a courtyard	central passage 4.10 m side passages ca. 2.50 m	40.26 m	3.30 m (= city wall)	width 8.94 m depth 18.87 m	Augustus
Colchester	Balkerne Gate	four-passage-gate	width about 18 m no gate building	central passage 5.10 m side passages 1.80 m	32.10 m	11.70 m	width 7.2 m depth 9 m	second century
Fréjus	Porte des Gaules	three-passage-gate	width of forecourt 50 m	central passage 5 m side passages 1.30 m			diameter 10 m	after 44 BC

Table 3.1 cont.

City	Name	Type	Dimensions gate building (without towers)	Width of passages	Total width	Depth of passages	Dimensions of towers	Date
Pompeii	Porta di Ercolano	three-passage-gate	width 14 m	central passage 4.4 m side passages 1.35 m	14 m	19 m		after 80 BC
Rome	Porta Appia	single gate	width 12.5 m depth 36 m	5 m	32 m	36 m	9.5 m by 9.5 m	271
Trier	Porta Nigra	double gate	width 15.91 m depth about 16 m	4.50 m	35.20 m	±16 m	width 10.05 m depth 21.47 m	175–200
Xanten	Burginatiumtor	double gate	width 12 m depth 7.33 m	4 m	26.68 m	7.33 m	width 7.34 m depth 10 m	115 AD

Source: Antalya: Schultze 327; Aosta: Schultze 295 and Taf. XIII; Autun: Schultze 306 and Taf. XIV (= figure 3.7); Colchester: Hull 18–19; Fréjus: Donnadieu 94 sqq (three-passage-gate); *contra* Schultze 292 (double gate); Pompeii: Schultze 287; Rome: Schultze 342; Giovagnoli 105; Trier: Cüppers 10; *contra* Schultze 335 and Taf. XVI (= figure 3.10) (slightly different dimensions); Xanten: Böcking 273 and plans of the APX.

Antiquity. Towers and gates are positioned on the city side of the walls, which limited defence considerably. These walls did not have a defensive function, but acted as a status symbol and clearly defined the city limits.

When there was no defensive function for the walls, was it then necessary to close the gates at night?

In Antiquity, little is written about gates and we do not know if there were stated opening- and closing-hours in times of peace. In the first century BC there was, however, a law according to Cicero that gates had to be closed at night.[809] However, he immediately mentions that even in times of war gates had to be opened in certain cases.[810] According to Gross, gates were closed at night; travellers had to take bedding and cooking utensils with them, in case they were forced to stay the night outside the city walls in the open or when places in inns were not available.[811]

However, there is evidence that gates were open at night.

At first, there were gates which, as mentioned earlier, *could* not be closed, because they did not have doors and/or portcullises. There were gates of this type in the Hellenistic provinces, especially as for instance in Antalya. The passages of these gates could be attended by soldiers, guards or nightwatchmen, but evidence of this is nowhere to be found. Aelius Aristides writes that he is leaving the little city of Cyme (in the neighbourhood of Pergamum) at midnight through the gate.[812]

A second argument for the opening of gates is given by the *Lex Julia Municipalis*. In this law it is stated that wagon traffic is forbidden by day, except triumphing emperors, Vestal Virgins, building traffic, wagons for the removal of garbage and excrement, and traffic leaving the city. The majority of large-scale traffic flow, for the supply of cities, had to take place by night, and according to this law gates had to be open at night. Nevertheless, it is not certain if the *Lex Julia Municipalis* was strictly adhered to everywhere in the Empire. For small, old cities in Italy, the law had some use, but in the later planned *coloniae* with their wide streets like Xanten it should not have been necessary to restrict wagon traffic by day.

A third piece of evidence is revealed by the gate-keepers.

In Latin, there is no word for 'gate-keeper'. The word *portarius* first occurs at the end of Antiquity, but the *Thesaurus Linguae Latinae* mentions only Bible passages with this word, meaning 'monastery gate-keeper'.[813] Besides that, there are the words *ianitor, ostiarius* and *custos*,[814] but these words refer to guards of private houses and other buildings, and not of city gates. In the Greek language, there is definitely a word for gate-keeper, πυλωρός, and even for gate-keeper's lodge, πυλώριον.[815] In the city-state society of Classical Greece, nearly all cities were surrounded by walls. After the Classical Greek period, the πυλωρός became a rarity.[816] Not only in literature, but also in archaeological and epigraphical sources, there is only very scarce evidence of

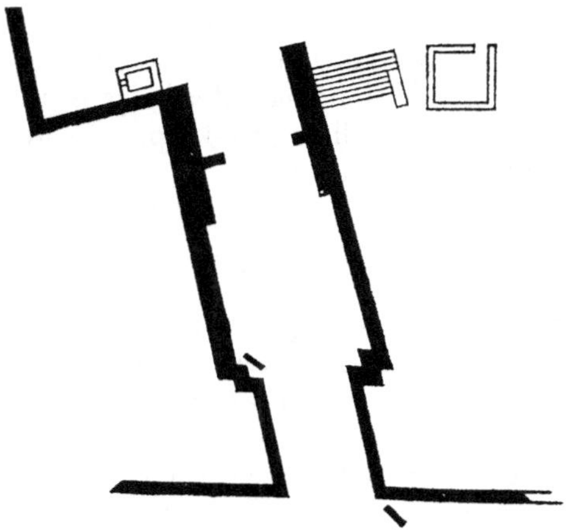

Figure 3.13 Pompeii, Porta di Stabia: right of the steps the gate-keeper's lodge. Overbeck 50. Leipzig: W. Engelmann.

gate-keepers. Mau, Etienne and Overbeck mention a gate-keeper's lodge next to the passage of Porta di Stabia (Fig. 3.13).[817]

The beginning of Pompeii's history was during a period of city-states and narrow gates with a high military importance. We are not sure if gate-keepers still closed gates in the Empire, certainly not when the *Lex Julia Municipalis* was also in force in Pompeii.

What would have been the task of a potential gate-keeper? The common image of this is twofold: the opening and closing of gates and the control of passing traffic. This is the image of medieval gate-keepers, who had to tend to the security in the city. In this period, every city had a wall with gates. In the Roman Empire, however, the situation was quite different. There were many open cities and those with walls had to open the gates at night for passing traffic – when the *Lex Julia Municipalis* was in force. So not only by day, but also by night, cities had to be accessible; the safety aspect was less important than closing the cities. Rich people had their own guards for their private houses and used a slave as door-keeper. Poorer people without slaves were less attractive for thieves and other criminals. Further, every city of course had a team of nightwatchmen, patrolling the streets at night.

The control of passengers at the gates by gate-keepers could also take place by day, as well as by night when the gates could be opened at night according to the *Lex Julia Municipalis*. It is not implausible that there was control of traffic entering and leaving, including the control of traffic in the case of single gates. Nevertheless, evidence of this has not been found. In Lyons, a toll was levied at the city gates, but this was done by tax officials who had nothing to do with the supervision of city gates. Control could have been done by the local guards, the *vigiles*. Matthews assumes that *vigiles* were placed on the approach roads to Rome and at the gates of the Aurelian Wall to block forbidden traffic.[818] The lodge at Porta di Stabia would also have been the place for a traffic-control official, because this lodge is situated within the city wall; a lodge for a traffic controlling gate-keeper would have been expected more outside the wall, on the land side.

Connection between city gates and suburbs

Is there a connection between the number of passages in a city gate and the existence of suburbs? To answer this question is not easy, because most city gates have disappeared and sources are scarce. So far as we know, there was no prescribed number of passages for the various situations. Existing gates were not changed much to give more space for traffic, even when the situation caused congestion. In some special cases, it was possible to replace an old gate with a new one with more passages; in Late Antiquity, it could happen that the number of passages diminished. Honorius bricked up some passages in the Aurelian Wall.

In the case of planned *coloniae*, a connection is sometimes visible between double gates on the one hand and three- and four-passage-gates on the other hand and the existence of suburbs. Cemeteries are found everywhere in the neighbourhood of every type of gate, in old and new cities. Some cases and situations will now be discussed. These are gates belonging to *coloniae* where information is available concerning their situation and building, and the presence of extramural buildings on their land side.

Pompeii: Porta Marina and Porta di Ercolano

Porta Marina, together with Porta di Ercolano, belongs to the 'newer' generation of city gates constructed in Pompeii. The gate has two passages: a large one for wagons and animals, and a smaller one for pedestrians. Overbeck and Mau date the construction of the gate to the second century BC, the period between the end of the Second Punic War and the Social War (202-90).[819] Immediately next to the gate, on the land side, there are baths, the Terme Suburbane, dating from the Empire.[820] The fact that these baths were supplied by sea water and the presence of a public-friendly double gate can explain the site of these baths.

Porta di Ercolano, dating from the first century BC,[821] is a three-passage-gate. The road spanned by this gate and known within the city as Via Consolare was the exit route to Herculaneum and Naples and one of the busiest in Pompeii.[822] The former gate, probably a single one, must have been a bottleneck. The immediate reason for building a new gate would have been war damage to the gate, caused by the siege of Sulla. Stone ammunition found here give evidence in that direction.[823] Maybe also the presence of large villas such as the Villa dei Misteri (second century BC, but renovated

Figure 3.14 Pompeii, Porta di Ercolano in its original state. Overbeck opposite 42. Leipzig: W. Engelmann.

60 BC)[824] was a reason to build a gate with more passages; the rich owners of these villas would have had the aim of building a more beautiful and imposing gate complex (Fig. 3.14).

City gates, forecourts and also the approach roads were part of the city and could not be occupied by private individuals alone, away from their public function. Private building and occupation without permission of the city government was not allowed.[825] Perhaps illegal building was also forbidden in order to diminish traffic flow as much as possible past the gates.

Cologne: Porta Paphia and other gates

In 50 AD, Cologne became a *colonia* and the planning and construction of the city wall was started immediately. This wall surrounded the former Germanic settlement Oppidum Ubiorum. When the surrounded area was filled up with houses and other buildings, workshops like potteries with a fire hazard were removed to the approach routes outside the walls.[826] Pottery, in the pre-colonial period already situated on the west side of the settlement, was in the Flavian era already entirely situated outside the west wall. There were also houses outside this gate, probably belonging to owners of the potteries.[827]

The wall contained, apart from the harbour gates, three important three-passage-gates (one on each land side), a double gate on the west side and two single gates on the west and south sides. The middle of the three harbour gates on the river side, called Markt- or Mars Tor, was also perhaps a gate with two passages.[828]

Of all these gates we have the most information on the main three-passage-gate on the north side, with its medieval name Porta Paphia. It was a three-passage-gate like Porta Praetoria in Aosta, with a large central passage and two smaller pedestrian passages, flanked by two towers.[829]

Looking at the plan (see p. 43, Fig. 2.1), it can be seen that most of the workshops are situated outside the three-passage-gates on the north, west and south sides of the wall. One can find a smaller concentration at the double gate on the west side. This gate was not a double gate like Porta Nigra in Trier, but was like Porta Marina in Pompeii: one high passage for wagons and horsemen, and a smaller one for pedestrians.[830] So we see here the opposite situation to that in Pompeii: after the initial planning and construction of the walls with the main gates spanning over *cardo* and *decumanus*, the extramural suburbs were then developed. For a good functioning traffic policy, the local government chose three-passage-gates.

Xanten: Burginatiumtor and other gates

The start of the building of the city wall at Xanten is dated, by means of dendrochronology, to the year 106,[831] after the acquisition of the status of *colonia* between 98 and 105.[832] Before that time there was already a settlement inside the projected walls as in Cologne, maybe the capital of the Germanic tribe the Cugerni. After the foundation of the *colonia* with its walls, the city area was built up. The construction of Burginatiumtor is dated, according to the discovery of coins, to around 115.[833]

Outside the gate, some foundations of a building have been found. If we assume that there is no further suburb of any significance,[834] this could explain why Burginatiumtor is a double and not a three- or four-passage-gate. The absence of a suburb of any importance did not necessitate special pedestrian passages in the gate. Traffic passing the gate was probably mostly long distance.

Figure 3.15 shows the south-east Veterator, spanning over the *cardo* leading to the military settlement Castra Vetera. The shaded reconstruction shows the plan of a double gate such as Burginatiumtor, but it is very disputable; only a small (black) part is excavated. It is remarkable that in most publications the plans of Colonia Ulpia Trajana show Veterator as a three-passage-gate; see for example Map IV (p. xviii).[835]

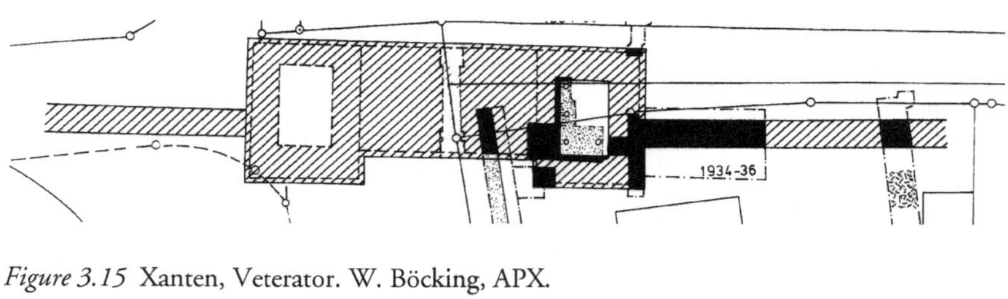

Figure 3.15 Xanten, Veterator. W. Böcking, APX.

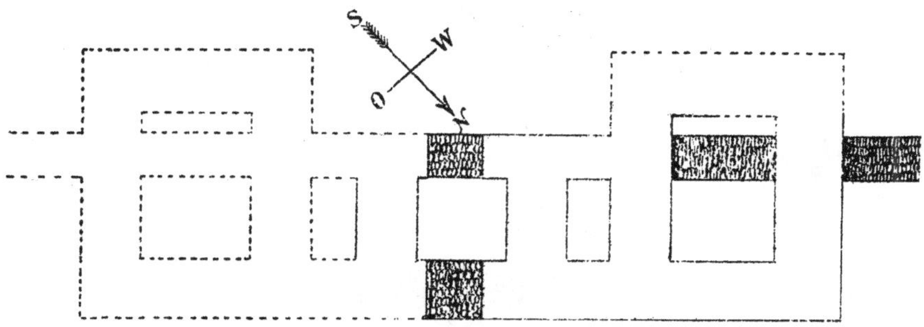

Figure 3.16 Xanten, Maastor. Lehner 182. Mainz: von Zabern.

The third main gate, Maastor, spanning the *decumanus* on the south-west side, had already been researched around 1900; according to Lehner and Bechert, it is a three-passage-gate (Fig. 3.16). The fact that towers are built over the wall indicates that there were two building phases, of which the second phase could be dated after a fire in the second half of the second century AD.[836] This fire could be the result of an invasion of the Chaukes in Belgica in the years 172–174; Didius Julianus, emperor in 193, expelled these barbarians.[837] If we follow the rule that in the case of three- and four-passage-gates there were suburbs, one can expect extramural buildings here; it was pointless and unnecessary to build three-passage-gates for pedestrians when there was hardly any pedestrian traffic.

Some other cities

Trier did not have suburbs at all[838] outside the second-century city wall and the gates were situated at a distance from the built-up area of the city (Fig. 3.17);[839] they were planned and constructed as double gates. Besides Porta Nigra, there was Porta Media, Porta Alba and Porta Inclyta. A fifth gate, part of the amphitheatre, contains three passages, but it would not be correct to qualify this gate as a three-passage-gate.[840]

Colchester had a four-passage-gate on its west side, Balkerne Gate. In front of the gate extramural buildings are found, including a temple. On the north-west side there were also extramural buildings; so here archaeologists assume a three-passage-gate. On these west and north-west sides there was a concentration of not only temples but also kilns (Fig. 3.18 and 3.19).[841] The other gates do not have buildings in front of them and are more simply constructed, with fewer passages.[842]

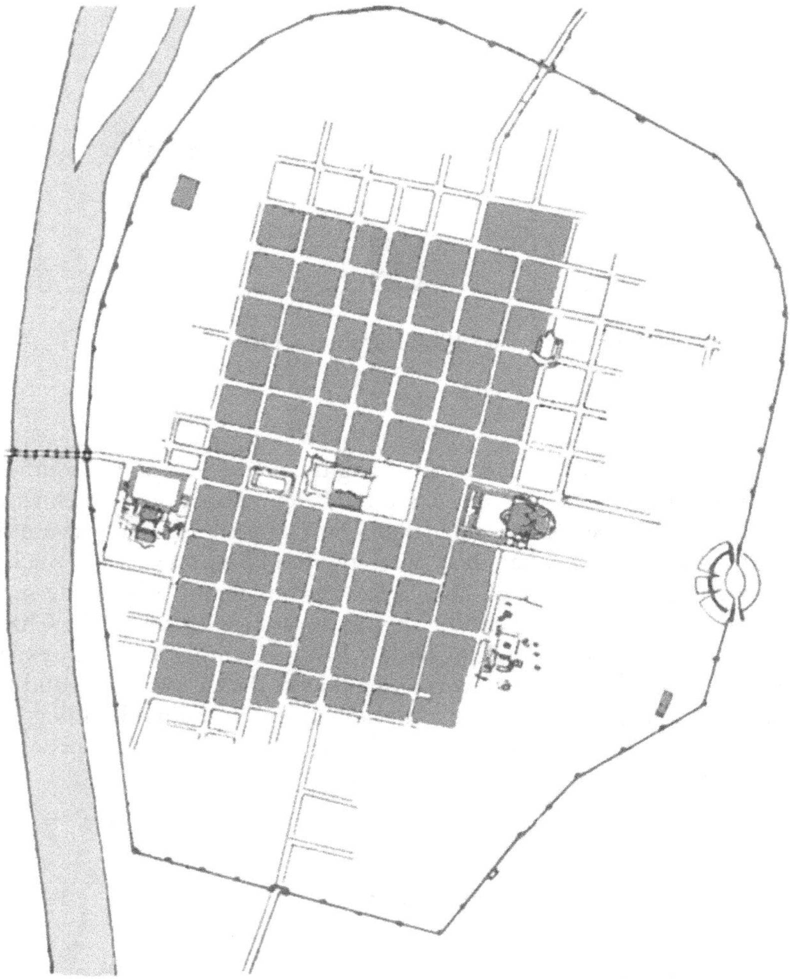

Figure 3.17 Trier, plan. Picture: C.R. van Tilburg.

Tongeren was a Roman city, but not a *colonia*. It was a city planned with a chessboard pattern (Fig. 3.20). At the beginning of the second century AD, the surrounding wall was erected, including five gates; only one of them, the west gate to Cassel, was constructed as a double gate. The other gates were single, in towers.[843] The reason why this method of gate construction was chosen was probably the relatively large number of gates (five or maybe six) and because the construction of larger-scaled gate complexes would have been too expensive. Moreover, the city was smaller than Xanten or Cologne. It is probable that these considerations

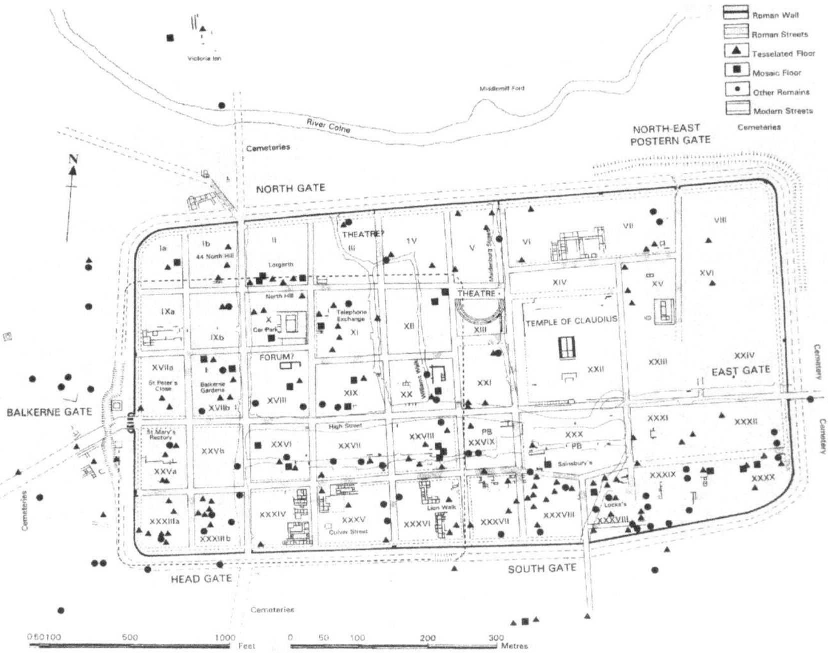

Figure 3.18 Colchester, plan. Wacher 1997, 115. London: Routledge.

were the reason to construct only one gate on the east side, for the exit route to Cologne. The former road ended at the wall.[844] The new gate spanned a new *decumanus* (removed to the north) which also connected with the double gate to Cassel; so the former *decumanus* lost its function as a thoroughfare.[845] In Tongeren we also see, in the case of two gates, the occurrence of the 'bird's foot': two or more roads coming together outside the gate.[846]

Along the road to Bavay, a large complex of granaries has been found. This complex was situated outside the city wall and was in the first instance older than the city wall (first century AD). The complex is orientated towards the road, not towards the city, and so a *statio* is assumed, which has nothing to do with the city.[847] The road from Cassel runs alongside this complex. So we have extramural buildings, but there was probably no busy traffic to the city of Tongeren similar to the industrial traffic of Cologne.

In *Rome*, we find a situation similar to Trier. The Aurelian Wall was erected on a large bend around the existing, older buildings – with gardens and green areas between the old city and the wall – and the gates are all single or double, with no suburbs of any importance.[848]

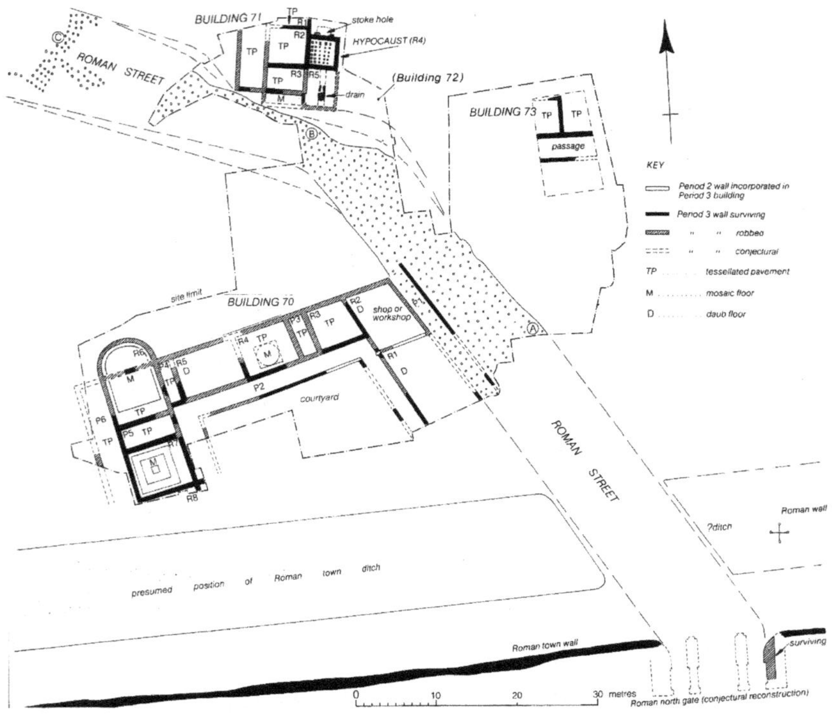

Figure 3.19 Colchester, north-west gate and suburb. Wacher 1997, 129. London: Routledge.

Summary and conclusion

The image we have of Roman cities, surrounded by a wall, with gates closed by gate-keepers at night, is a medieval image. Many Roman cities did not have fortifications at all; they were open cities. Cities with a wall were probably also open at night.

A clear indication of the existence of gate-keepers has not been found. Literary sources do not speak of gate-keepers and archaeological sources are very scarce. Probably there were gate-keepers in areas and during times that cities were more or less independent, like the city-states of Classical Greece. But it is doubtful if the profession of gate-keeper existed in the Empire. When, according to the *Lex Julia Municipalis*, gates also had to be open at night there was obviously no need for the existence of gate-keepers to close the gates.

Extramural buildings could cause a lot of traffic through the gates. In older cities, like Pompeii, gates could be simple and narrow. Widening of gates for a better management of traffic did not in fact take place; one obviously accepted the traffic congestion. In cases of damage or destruction to a gate, for example by war, one could build a new, wider one. The question of how many passages a gate could contain depended on the existence of suburbs or (in the case of *coloniae*) future suburbs. In Pompeii, there were already extramural buildings outside Porta di Ercolano; in Cologne, the gates were built first and extramural building could be developed at a later stage; workshops causing both nuisance and fire hazard were formerly situated inside the walls.

In the case of double gates, extramural buildings were scarce or completely absent. In *coloniae*, traffic flow seems to have been planned when the city was planned, so there was a certain traffic policy in place.

Crowds in the city

Introduction

Compared with our cities nowadays, cities in Antiquity were smaller and more concentrated and so the traffic flow was also smaller. In the preceding chapters it was stated that the interurban traffic by land was of relatively small volume. The heaviest traffic was the goods transport between city and countryside. Between the city and suburbs traffic could also be considerable; during the planning of the city, the city gates could be adapted for traffic flow. Traffic *inside* the city was lighter and consisted of people working away from home – especially officials – and goods transport from the approach roads and harbours to the distribution centres and the shops.

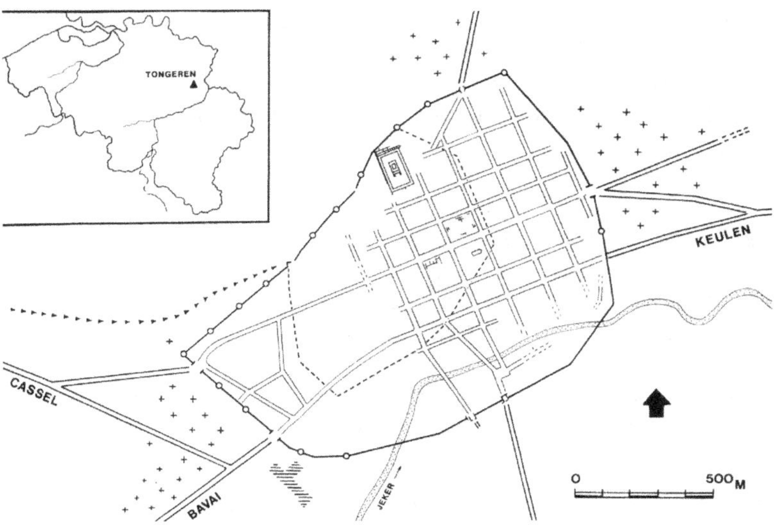

Figure 3.20 Tongeren, plan. Mertens and Vanvinckenroye 7. Tongeren: Publicaties van het Provinciaal Gallo-Romeins Museum Tongeren.

Besides that, the streets were full of other crowds; the shopping public and people who were on the road for their profession – street-traders, pedlars, market-merchants, etc. – also caused crowds in the streets, especially in the older cities. In these cities, the traffic congestion must have been the biggest problem and the local governments needed to take measures to keep charge of the urban logistics.

Crowds in Rome

There is no doubt that Rome – more than any other city in the Empire – suffered from its urban traffic chaos. There were three causes. First, the irregular infrastructure caused by disorganised city planning[849] and uncontrolled expansion in the second and first century. The realisation of a Hippodamic street pattern was also a problem caused by the winding Tiber and the presence of hills. The relatively flat Campus Martius had a Hippodamic structure; outside Campus Martius, the biggest buildings – Circus Maximus, Baths of Trajan, Baths of Caracalla and Baths of Diocletian – have a north-west to south-east orientation. This orientation also pinpoints the *fora* and palaces on the Palatine Hill, albeit not exactly parallel. The second cause was the large population density[850] and the third a warm climate, which made it possible to live on the street for a large part of the time.

It is not possible to say exactly how many inhabitants Rome had, but probably more than a million at the beginning of the Empire.[851] The surface-area of Rome was smaller than that of Alexandria and Antioch, each with about 500,000 inhabitants.[852] The population density of Rome was therefore much higher than that of the other two cities. Here follow some citations on the crowds in Rome. Below Horace writes about heavily loaded wagons and animals on the street:

Festinat calidus mulis gerulisque redemptor,
torquet nunc lapidem, nunc ingens machina tignum,
tristia robustis luctantur funera plaustris,
hac rabiosa fugit canis, hac lutulenta ruit sus;
. . .
tu me inter strepitus nocturnos atque diurnos
vis canere et contracta sequi vestigia vatum?

In hot haste rushes a contractor with mules and porters; a huge crane is hoisting now a stone and now a beam; mournful funerals jostle massive wagons; this way runs a mad dog: that way rushes a mud-bespattered sow. . . . Would you wish me, amid noises by night and noises by day, to sing and pursue the minstrels' narrow pathway?

(Hor. *Ep.* 2.2.72–80)

Also Juvenal has complaints about traffic noise:

raedarum transitus arto
vicorum in flexu et stantis convicia mandrae
eripient somnum Druso vitulisque marinis.
si vocat officium, turba cedente vehetur
dives et ingenti curret super ora Liburna
atque obiter leget aut scribet vel dormiet intus;
namque facit somnum clausa lectica fenestra.
ante tamen veniet: nobis properantibus opstat
unda prior, magno populus premit agmine lumbos
qui sequitur; ferit hic cubito, ferit assere duro
alter, at hic tignum capiti incutit, ille metretam.
pinguia crura luto, planta mox undique magna
calcor, et in digito clavus mihi militis haeret.
nonne vides quanto celebretur sportula fumo?
centum convivae, sequitur sua quemque culina.
Corbulo vix ferret tot vasa ingentia, tot res
inpositas capiti, quas recto vertice portat
servulus infelix et cursu ventilat ignem.
scinduntur tunicae sartae modo, longa coruscat
serraco veniente abies, atque altera pinum
plaustra vehunt; nutant alte populoque minantur.
nam si procubuit qui saxa Ligustica portat
axis et eversum fudit super agmina montem,
quid superest de corporibus? quis membra, quis ossa
invenit? obtritum vulgi perit omne cadaver
more animae.

The crossing of wagons in the narrow winding streets, the slanging of drovers when brought to a stand, would make sleep impossible for a Drusus[853] – or a sea-calf. When the rich man has a call of social duty, the mob makes way for him as he is borne swiftly over their heads in a huge Liburnian car. He writes or reads or sleeps inside as he goes along, for the closed window of the litter induces slumber. Yet he will arrive before us; hurry as we may, we are blocked by a surging crowd in front, and by a dense mass of people pressing in on us from behind: one man digs an elbow into me, another a hard sedan-pole; one bangs a beam, another a wine-cask, against my head.

> My legs are beplastered with mud; soon huge feet trample on me from every side, and a soldier plants his hobnails firmly on my toe.
>
> See now the smoke rising from that crowd which hurries as if to a dole: there are a hundred guests, each followed by a kitchener of his own. Corbulo himself could scarce bear the weight of all the big vessels and other gear which that poor little slave is carrying with head erect, fanning the flame as he runs along. Newly-patched tunics are torn in two; up comes a huge fir-log swaying on a wagon, and then a second dray carrying a whole pine-tree; they tower aloft and threaten the people. For if that axle with its load of Ligurian marble breaks down, and pours an overturned mountain on to the crowd, what is left of their bodies? Who can identify the limbs, who the bones? The poor man's crushed corpse wholly disappears, just like his soul.
>
> (Juv. 3.236–61)

And Seneca:

> *Cogitato, in hac civitate, in qua turba per latissima itinera sine intermissione defluens eliditur, quotiens aliquid obstitit, quod cursum eius velut torrentis rapidi moraretur, in qua tribus eodem tempore theatris caveae postulantur.*
>
> Consider this city, in which the throng that streams ceaselessly through its widest streets is crushed to pieces whenever anything gets in the way to check its course as it streams like a rushing torrent, – this city in which the seating space of three theatres is required at one time.
>
> (Sen. *Cl.* 1.6.1)

So in Rome there was enormous traffic congestion. Not all wagons were forbidden by day; the *Lex Julia Municipalis* permitted wagons to drive around the city during the day, loaded for example with building material; Juvenal cited wood and stone (marble) transport. The danger of being crushed was not imaginary; the Late Republican jurist Alfenus Varus describes a dispute between two mule wagons which resulted in a boy being crushed.[854]

The traffic situation deteriorated even further with the building of imperial *fora*, creating vast areas exclusively for pedestrians[855] and the annexation of parts of streets by shop- and bar-keepers. In 92, Domitian ordered that streets be kept free of obstacles.[856] These and other rules and regulations, being attempts to control the traffic chaos somewhat, will be discussed below, pp. 127–36.

Crowds in other cities

In many cities in Italy there must also have been much traffic congestion. Claudius enacted a law decreeing that one could only go into a city on foot or by sedan.[857] Perhaps this law was enacted owing to the fact that Cn. Domitius Ahenobarbus, Nero's father, ran over and killed a boy.[858] A city like Pompeii was set up for the most part Hippodamic (like a chessboard), but most streets were only suitable for one-way traffic and city gates were usually narrow ones. Obeying Claudius' law, one had to lead the horses and asses by the reins through the cities. Gesemann is uncertain if Claudius' law was also binding for through-traffic using streets such as Via Stabiana and Via dell'Abbondanza, which had sidewalks and enough space.[859]

Cities outside Italy too, especially the metropoles Alexandria, Antioch, Pergamum, Ephesus, Carthage and, in Late Antiquity, Constantinople, must have had huge crowds, but because of better planning, Rome-like situations did not occur. The presence of sidewalks and colonnades could give pedestrians a certain amount of protection against wagons. In Alexandria, the size of the population and the facilities of the city seem to have been in balance. Achilles Tatius praises the city:

> *καὶ εἰ μὲν εἰς τὴν πόλιν ἀπεῖδον, ἠπίστουν εἰ πληρώσει τις δῆμος αὐτὴν ἀνδρῶν, εἰ δὲ εἰς τὸν δῆμον ἐθεασάμην, ἐθαύμαζον, εἰ χωρήσει τις αὐτὸν πόλις. τοιαύτη τις ἦν ἰσότητος τρυτάνη.*

> Looking at the city, I doubted whether any race of men could ever fill it; looking at the inhabitants, I wondered whether any city could ever be found large enough to hold them all. The balance seemed exactly even.
>
> (Ach. Tat. 5.1.6)

The fact that cities could also be busy in Late Antiquity has already been mentioned above, p. 48, when Jerome complained about the crowds. The upsurge of Christianity had caused new pilgrim traffic in Jerusalem.[860] Ausonius, a poet living in the fourth century, describes the crowds in Bordeaux:

> *Nam populi coetus et compita sordida rixis*
> *fastidientes cernimus*
> *angustas fervere vias et congrege volgo*
> *nomen plateas perdere.*
> *turbida congestis referitur vocibus echo:*
> *'Tene, feri, duc, da, cave!'*
> *sus lutulenta fugit, rabidus canis impete saevo*
> *et impares plaustro boves.*

> For I am weary at the sight of throngs of people, the vulgar brawls at the cross-roads, the narrow lanes a-swarm, and the broadways belying their name for the rabble herded there. Confused Echo resounds with a babel of cries: 'Hold!' – 'Strike!' – 'Lead!' – 'Give!' – 'Look out!' Here is a mucky sow in flight, there a mad dog in fell career, there oxen too weak for the wagon.
>
> (Aus. *Ep.* 6.19–26)

It is not certain if Bordeaux was in reality such a busy city as Ausonius is describing here. There must have been busy traffic, but uncontrolled Roman-like situations are doubtful. Bordeaux was a splendid city, according to Ausonius, accommodated with wide streets and squares.[861] Maybe Ausonius wanted to praise Bordeaux with this citation, partially derived from Horace.

Traffic regulations and traffic aggression

One would expect that there must have been traffic regulations for well-functioning traffic, especially in a city. Neither in literature, nor in archaeology, nor in epigraphy is there information on this theme. Taking into consideration the aforementioned citations, there was probably no talk of it. There is an assumption that Roman traffic drove on the left. This is supported by British archaeologists from research on wheel ruts in a limestone quarry at Blunsdon Ridge, Swindon (between Oxford and Bristol).[862]

The lack of regulations made it possible and even inevitable that drivers drove, where possible, in the middle of the road, on condition, of course, that this behaviour did not damage their own or others' wagons or property, or cause fatal or other accidents. Slanging-matches must have been the rule rather than the exception and coachmen and carters had the reputation for insolent language up until the twentieth century. Lictors had the task of beating all road-users getting in the way, in and outside the city, with their *fasces*, which outside the city were augmented by an axe.

One can assume that the 'man in the street' could also make use of a lictor-like person. In the Baths of the Coachmen in Ostia a mosaic can be seen (Fig. 3.21) where a wagon, probably a *rheda*, is preceded by a man with a whip. Being a harbour city with many harbour workers, porters and merchants, Ostia must have been busy, but the rental of a wagon with the destination Rome must also have been a possibility. It was probably busy in the streets before Rome's *pomerium*; after the passage of the *pomerium*, when the *Lex Julia Municipalis* was in force, it was forbidden to drive a *rheda* by day (unless the driver had permission). Nevertheless, this solution was not the perfect one. The *rheda* could not overtake the walking whip-carrier and therefore could not drive faster than a pedestrian. So the question is whether such a situation was normal.

Figure 3.21 Ostia, mosaic in the Baths of the Coachmen. Photo: C.R. van Tilburg.

Besides the abundance of traffic and – as far as we know – the absence of traffic regulations, the crowds were worsened by the lack of good traffic signs, obstructions like market stands, deficient city gates, fountains, etc. In the case of special events like the passing of an army, a festival, a religious or triumphal procession, the situation was even worse.

Summary and conclusion

In contrast to interurban routes, where traffic congestion was scarce, the crowds in the cities were considerable. The worst logistical problems took place in older, organically developed cities in Italy and other (Mediterranean) parts of the Empire. There is no doubt that the problems in Rome were the worst. Measures and laws could not prevent the fact that traffic chaos has been exposed in literature, which scarcely mentions traffic and its associated congestion outside the city.

In older cities, there was for road-users a maxim of 'survival of the fittest', especially in Rome. Whoever, in spite of the crowd, did not get away fast enough from an approaching lictor or wagon ran the risk of a slanging-match or physical violence. In later cities, adequately built and planned with a chessboard pattern, driving-traffic had more space. In a *colonia* like Xanten, where streets were wide enough to permit parked wagons without being an obstacle for passing vehicles, the chaotic situations characteristic of Italian cities did not occur.

Conclusion of chapter 3

In this chapter, an attempt has been made to answer the question whether there was traffic congestion in the Roman Empire. There was no acute traffic congestion between cities on long-distance roads and intersections, like we see nowadays. Leaving the city and suburbs, one came into the thinly populated or uninhabited countryside, where only toll points could cause delay – apart from accidents, breakdown or crime. Delay could also, of course, be caused by marching armies, heavy transport or a travelling emperor, but these contingencies were not a daily problem. We can state that there was not a structural problem on the long-distance routes.

On a local level, there were two categories of city: planned cities where one reckoned with as good a traffic flow as possible and older, organically grown cities. If surrounded by a wall, planned cities were accommodated with city gates, able to sustain heavy traffic flow. One could also decide where suburbs, mainly workshops and graves, could be sited. Traffic congestion was thus limited.

The older (Mediterranean) cities did not only have narrow streets, but usually also narrow, single city gates, constructed at a time when cities had to be fortified against enemies; the importance of traffic succumbed to the importance of defence. In these cities traffic congestion was a problem, not only around city gates, but also inside the walls. To maintain accessibility wheeled traffic was forbidden by day, but because significant groups of wheeled traffic had permission to drive in the cities, situations with a nuisance and danger element remained a continual problem, aggravated by the absence of traffic signs and regulations. Rome, the vast metropolis with more than a million inhabitants, without a planned street system and with a very high population density, had traffic problems, probably not a lot different from modern cities nowadays. Congestion, stench and noise were part of daily life. That we read few written complaints is probably caused by the fact that most authors lived far from the problems of the city.

There remains one question: what means did local and central government use to keep the situation under control or, where possible, to provide a solution? This question will be discussed in the next and final chapter.

4

TRAFFIC POLICY

Nowadays, governments try to diminish traffic congestion by taking – mostly unpopular – measures. Were such measures to improve the traffic situation in the Roman Empire also taken by the government of the day? In this chapter that question will be discussed.

Caesar, and other emperors, enacted laws and orders, in an attempt to improve the traffic situation. These laws were in force in Rome, where the traffic situation was the worst. Other cities in Italy were also subject to these laws. As well as that, they could set up and enforce laws themselves. The question remains whether everybody obeyed these laws. Implementing them laws must have been a difficult matter.

The placing of barricades in the streets had more effect. Excavations in Pompeii show that wheeled traffic could not drive everywhere, but was forced to use fixed routes. Elsewhere, it was impossible for wagons to pass, because the streets were completely blocked by stone edges. Such obstructions must have been a normal occurrence in older cities. In newer cities too, including the *coloniae*, fixed routes were in use.

In this chapter the opportunities not taken by governments will be discussed. Many solutions, now used by governments today, must have been obvious in Antiquity, but they were not put into practice. Measures taken by the government were limited to freedom of movement; extension of the infrastructure, nowadays seen as the solution to traffic congestion, did not take place.

Legislation

Introduction

In the previous chapters a number of laws and provisions have already been mentioned, which were established to reduce traffic hindrance. As well as the *Lex Julia Municipalis* there were also other city laws (*leges municipales*), established by individual cities. They are related to the organisational structure of cities and give among others things a picture of spatial planning. Apart from the *leges municipales*, all related to cities and boroughs in southern Italy and Spain, there were also imperial provisions aimed at reducing traffic chaos in cities, particularly Rome. In this section, some of these *leges municipales* related to spatial planning and traffic maintenance will be discussed.

Early traffic legislation

In chapter 1 (p. 27) it has been stated that as early as in 450 BC, in the Law of the Twelve Tables, the stipulation was made that through-roads, *viae*, had to have a minimum width of 8 feet and a minimum of 16 feet at bends. With this provision, it is clear that many roads were obviously too small. Whether there was already traffic congestion in that archaic time is implausible; the provision merely made it possible for wagons to pass and overtake each other.

Also in chapter 1 (pp. 6, 34), the *Lex Sempronia viaria* of C. Gracchus has been mentioned. As well as this law, we know of a law of C. Scribonius Curio, a *lex viaria* with little success[863] and a *Lex Visellia*.[864] These laws concern road building and not traffic movement control.

The **Lex Julia Municipalis**

The *Lex Julia Municipalis* or *Tabula Heracleensis* is the most famous law text concerning traffic. The text was found in 1732, in the neighbourhood of the southern Italian city of Heraclea (Gulf of Taranto) and contains provisions in force in Rome, but we must assume that they were also valid in Heraclea.[865] According to Mommsen, the *Lex Julia Municipalis* was in force in Padua, because there is talk of M. Junius Sabinus, a *quattuorvir aedilicia potestate e lege Julia municipali.*[866]

There is disagreement about the year of introduction. Van Binnebeke mentions 60 BC;[867] Homo, Marquardt and Sonnabend have 45 BC, the year also mentioned by the *CIL*; Legras mentions the years after 90 BC and, finally, Nap has 65 BC.[868]

Here follows the text and translation of the citation of the *Lex Julia Municipalis* concerning the prohibition of wagon traffic in Rome by day:

> *Quae viae in u(rbem) R(omam) sunt erunt intra ea loca, ubi continenti habitabitur, nequis inieis vieis post k(alendas) Ianuar(ias) | primas plostrum interdiu post solem ortum, neve ante horam X diei ducito agito, nisi quod aedium | sacrarum deorum inmortalium caussa aedificandarum, operisve publice faciumdei causa, adv< e >hei porta|ri oportebit, aut quod ex urbe ex ve ieis loceis earum rerum, quae publice demolienda< e > loca< tae > erunt, publi|ce ex portarei oportebit, et quarum rerum caussa plostra h(ac) l(ege) certeis hominibus certeis de causeis agere | ducere licebit.*

Quibus diebus virgines Vestales, regem sacrorum, flamines plostreis in urbe sacrorum publicorum p(opuli) R(omani) caussa | vehi oportebit, quaeque plostra triumphi caussa, quodie quisque triumphabit, ducei oportebit, quaeque | plostra ludorum, quei Romae <p(ropius) p(assus) M> publice feient, inve pompam ludeis circiensibus ducei agei opus | erit, quo minus earum rerum caussa eisque diebus plostra interdiu in urbe ducantur agantur, e(ius) h(ac) l(ege) n(ihil) r(ogatur).

Quae plostra noctu in urbem inducta erunt, quominus ea plostra inania aut stercoris exportandei caussa, | post solem ortum h(oris) X diei bubus iumenteis veiuncta in u(rbe) R(oma) et ab u(rbe) R(oma) p(assus) M esse liceat, e(ius) h(ac) l(ege) n(ihil) r(ogatur).

In the roads which are or shall be within the city of Rome, or within the limit of continuous habitation, no person, after the first day of January next following, shall be allowed in the daytime, after sunrise or before the tenth hour[869] of the day, to lead or drive any heavy wagon; except where it shall be requisite, for the sake of building the sacred temples of the immortal gods or carrying out some public work, to draw or convey material into the city, or where, in pursuance of a contract for the demolition of buildings, it shall be requisite for public ends to carry material out of the city or away from such places, and in cases and for objects for the which it shall be lawful for specified persons and for specified causes to lead or drive such wagons.

On all days when the Vestal Virgins, the rex sacrorum, and the flamins shall be required to ride in wagons in the city for the sake of the public sacrifices of the Roman people, or when wagons shall be required for a triumphal procession on the day fixed for such triumph, or for games which shall be publicly celebrated within the city of Rome or within one mile of the city, or for the procession at the Ludi Circenses, for all such causes and on all such days it shall be lawful for wagons to be led or driven in the city in the daytime, notwithstanding any enactments in this law.

It shall be lawful for wagons, brought into the city by night, drawn by oxen or horses, if returning empty or conveying away refuse,[870] to be in the city of Rome or within one mile of the city after sunrise in the first ten hours of the day, notwithstanding any enactments in this law.

(*Lex Julia Municipalis* [*CIL* I^2 593] 56–67, tr. E.G. Hardy)

It is interesting that there is talk of Rome and of *ea loca, ubi continenti habitabitur*, a sort of suburb. By *urbs* Rome itself is meant, and the part that was situated inside the Servian Wall. In the course of the second century

BC, after the devastation of Carthage, the north part of this wall, however, disappeared and the *pomerium* was also moved. In 192 BC there were already extramural buildings around Porta Flumentana at the Tiber;[871] Varro also writes – in the first century BC – about extramural buildings at Porta Flumentana, where he also mentions a suburb: Aemilianae.[872] Aemilianae or Aemiliana was situated outside the Servian Wall, on the north side.[873] When the *Lex Julia Municipalis* became effective, the traffic situation in the suburbs, outside the Servian Wall, was in any case no less chaotic than within the (former) city wall.

The *Lex Julia Municipalis* permitted quite a lot of wagon traffic in the city; hence the complaints of several authors which were mentioned in the preceding chapter. Construction traffic could move by day, with the exception of traffic for private houses and also hire apartments, so-called *insulae*. Moreover there were a lot of empty wagons and wagons carrying rubble, waste and excrement.[874] The quantity of excrement must have been enormous; not only of people, but also of animals. The quantity and the hindrance of rubble, waste and excrement must have been so large that transport had to take place by day.

The question is whether the law was established to protect pedestrians against busy traffic. Rather, we must consider the possible separation of private traffic – inclusive of goods traffic, the supply of shops and goods delivery for private purposes – from the aforementioned government traffic. By day, government traffic would not in any case be impeded by private traffic.

Passenger and goods transport by wagon were thus the only traffic flow directly subject to restrictions under the *Lex Julia Municipalis*. The army and the *cursus publicus* (gold and money transports!) would have been exempt from this law, although the law does not mention this. Heavy transport and transport of large animals for circus and amphitheatre games were also government transport and therefore were also excluded from the restrictions of the *Lex Julia Municipalis*.

It is not certain if individuals and private entrepreneurs – farmers from the *suburbium* come to mind – effectively experienced much hindrance with the introduction of the *Lex Julia Municipalis*. Complaints concerning the introduction of the law are not known[875] and the question is whether there in fact were any. Individuals could rent a vehicle without many problems at the gates. The possession of a coach or wagon[876] in a city like Rome was more a disadvantage than an advantage in the crowded streets. Parking space was hard to find and most Romans were much too poor to buy a vehicle.[877] Traders and (market) merchants used smaller barrows or pack animals on which the *Lex Julia Municipalis* did not impose any restrictions. Considering the daily chaos, however, many transporters must have preferred to supply at night before the introduction of the *Lex Julia Municipalis*. Slave labour was available both by day and by night.

The Lex Coloniae Genetivae Juliae

We come across another citation concerning traffic in the *Lex Coloniae Genetivae Juliae*, a city law originating from the southern Spanish city of Urso. In this law the activities of various magistrates are described, including those of *IIviri* and aediles. As in every city law[878] the provision had been made that streets and pavements belonged to the public space and that individuals could not occupy it:

> *Quae viae publicae itineraque publica sunt fuerunt | intra eos fines, qui colon(iae) dati erunt, quicumq(ue) | limites quaeque viae quaeque itinera per eos a|gros sunt erunt fueruntve, eae viae eique limites | eaque itinera publica sunto.*
>
> Respecting public roads and footpaths within the boundaries assigned to the colony, all such thoroughfares, roads, and footpaths, which exist or shall exist or have existed in the said territories, shall be public property.
>
> (*Lex Coloniae Genetivae Juliae* [*CIL* I^2 594] LXXIIX, tr. E.G. Hardy)

The public character of the roads is again stressed in the *Lex Coloniae Genetivae Juliae* by the obligation to keep the roads free of obstacles:[879]

> *Qui limites decumanique*[880] *intra fines c(oloniae) G(enetivae) deducti facti|que erunt, quaecumque fossae limitales in eo agro erunt, | qui iussu C. Caesaris dict(atoris) imp(eratoris) et lege Antonia*[881] *senat(us)que | c(onsulto) pl(ebi)que sc(ito) ager datus atsignatus erit, ne quis limites | decumanosque opsaeptos neve quit immolitum neve | quit ibi opsaeptum habeto, neve eos arato, neve eis fossas | opturato neve opsaepito, quo minus suo itinere aqua | ire fluere possit. Si quis atversus ea quit fecerit, is in | res sing(ulas), quotienscumq(ue) fecerit, (sestertios mille) c(olonis) c(coloniae) G(enetivae) I(uliae) d(are) d(amnas) esto | eiusque pecun(iae) qui volet petitio p(ersecutio)q(ue) esto.*
>
> Respecting all boundary roads or cross roads, made or marked out within the territories of the *colonia* Genetiva, and all boundary ditches within the land, given and assigned by order of G. Caesar, dictator and imperator, and by the *Lex Antonia* and by decrees of senate and by plebiscite; no person shall have the said boundary roads or cross roads blocked up, nor have any heaps or obstructions

> therein, nor plough over the same, nor block nor obstruct the said ditches, whereby water may be hindered from running and flowing in its proper course. Any person acting in contravention of this shall for every several such act be condemned to pay to the colonists of the *colonia* Genetiva Julia 1,000 sesterces, and may be sued or prosecuted by any person at will for that amount.
>
> (*Lex Coloniae Genetivae Juliae* [*CILI*^{2} 594] CIIII, tr. E.G. Hardy)

The drift of these provisions is clear. Roads may not be blocked or occupied by individuals, and the traffic flow must be maintained. It is also forbidden to deface roads by cutting through with a plough or other tool.

What do *immolitum* and *opsaeptum* mean of which the *Lex Coloniae Genetivae Juliae* speaks? Goods causing blockages in the streets comes to mind (a situation prohibited by Domitian in Rome), but also for example a small barn or animal shed in a cul-de-sac. A cul-de-sac, however, also belongs to the public domain.[882] Furthermore, parked wagons come to mind. Parking problems are mentioned nowhere in literature; wagons were probably not left on the street when they were not in use, but in barns, gardens or promenades. In Pompeii, the House of Menander had a wagon shed; in Villa Regina in Boscoreale a garden was used.[883] If parked wagons were considered as *opsaepta*, a parking fine of 1000 sesterces would apply according to this law.

Traffic laws in the Empire

The restrictions of the *Lex Julia Municipalis* only related to wagons, vehicles with wheels. For riding and pack animals such as horses, mules and oxen restrictions were not imposed as far as we know. Because movements with these animals nevertheless caused trouble – there was, possibly, more need to use animals when the use of wagons was restricted – in the year 42 Claudius brought in a considerable reinforcement of the *Lex Julia Municipalis*. In the whole of Italy, not only was the use of wagons prohibited, but it was also forbidden to go through a city on horseback or by mule; only walking or by sedan:

> *Viatores ne per Italiae oppida nisi aut pedibus aut sella aut lectica transirent, monuit edicto.*
>
> He provided by an edict that travellers should not pass through the towns of Italy except on foot, or in a chair or litter.
>
> (Suet. *Cl.* 25)

Suetonius is speaking of *viatores*, travellers, not goods transport. Riding on animals obviously caused dangerous situations; the running over and killing of a boy by the father of Nero, Cn. Domitius Ahenobarbus, has already been mentioned in chapter 3. It must also be noted that this measure applied explicitly in Italy; outside Italy, in the cities, there was obviously space enough for horsemen possibly because the interests of pedestrians were considered less important. Besides that, not only horsemen but also animals could cause dangerous situations.[884]

Anyone approaching a city while riding an animal had to dismount and lead the animal by the reins. If there was no one supervising or there were no crowds, it must have been tempting to ride at walking pace in many cases. Strict observance of the law must have been impractical and later emperors were obliged to adhere to the *Lex Julia Municipalis* and Claudius' edict.[885]

Not only riding traffic, but also pedestrians were hindered by blocked streets and sidewalks. In Rome, shop- and bar-keepers took over parts of the street to carry out their business, so traffic chaos by day increased. In 92, Domitian passed a law prohibiting these activities. Martial's epigram 7.61 is cited here:

Abstulerat totam temerarius institor urbem
inque suo nullum limine limen erat.
Iussisti tenuis, Germanice, crescere vicos,
et modo quae fuerat semita, facta via est.
Nulla catenatis pila est praecincta lagonis
nec praetor medio cogitur ire luto,
stringitur in densa nec caeca novacula turba,
occupat aut totas nigra popina vias.
Tonsor, copo, cocus, lanius sua limina servant.
Nunc Roma est, super magna taberna fuit.

> The audacious shop owners had occupied the whole city; no doorsill kept within its own limits. You, Germanicus,[886] ordered the narrow streets to be wider, and what had recently been a track is now a street. No column is surrounded by chained flagons, and the praetor is not forced to plough through the mud. The Shaver is not drawn blindly in a dense crowd, nor does the dirty inn occupy the whole street. The barber, the inner, the cook, the butcher keep to their own limits. Now it is Rome, but it used to be a big shop.
>
> (Mart. 7.61)

The question is whether shopkeepers obeyed this law. This has already been mentioned in chapter 1 (p. 38), as well as the measures which aediles could take as regulations were broken. As far as that was concerned, the interests of the passers-by weighed more heavily than those of the shopkeepers.

Hadrian once again confirmed the *Lex Julia Municipalis* and Claudius' edict:

> *Vehicula cum ingentibus sarcinis urbem ingredi prohibuit. Sederi equos in civitatibus non sivit.*
>
> He forbade the entry into Rome of heavily laden wagons, and did not permit riding on horseback in cities.
>
> (*Historia Augusta, Hadrianus* 22.6)

The explicit declaration *cum ingentibus sarcinis* could indicate that wagons could only carry a modest load.[887] Heavy or wide loads could cause not only hindrance to other road-users, but also damage to houses or pavements.[888] There were also again too many mounted horsemen in the cities. Marcus Aurelius again decreed:

> *Idem Marcus sederi in civitatibus vetuit in equis sive vehiculis.*
>
> He forbade riding and driving within the limits of any city.
>
> (*Historia Augusta, Marcus Antoninus* 23.8)

Such a measure is very radical. Riding not only a horse is forbidden, but also on or in a wagon. The assumption is that this is only a confirmative of the *Lex Julia Municipalis*, where it is prohibited to drive a wagon by day, with the usual exceptions. Another possibility is that empty wagons can also drive by day in cities, as well as unridden horses. The consequence is, however, that the number of road-users increases and therefore leads to more crowds in the street. The doctor Galen, who visits a rich man in a suburb of a Roman city, mentions a place where people usually have to descend from their wagons, probably before the point where the city begins.[889]

It is not clear if the emperor himself was subjected to the *Lex Julia Municipalis* or other laws. If there was a triumphal procession, the *Lex Julia Municipalis* permitted the use of a wagon by day: *plostra triumphi caussa, quodie quisque triumphabit, ducei oportebit*. At any rate, Claudius used a wagon during his triumphal procession, where Messalina also sat on a *carpentum*.[890] From the reign of Septimius Severus, however, wagons became a privilege for the political elite. During his reign legates could drive by wagon where they first had to go on foot.

> *factum ut in vehiculo etiam legati sederent, qui ante pedibus ambulabant.*

> Legates, who had previously gone on foot, therafter rode in carriages.
> (*Historia Augusta, Severus* 2.7)

In 203, the *praefectus praetorio* Plautianus also drove by wagon[891] and the possession of an elegant wagon in which to travel became a privilege for the political upper class. Alexander Severus permitted senators to use silver-mounted *carrucae* and *rhedae*:

> (Alexander Severus) *Leges innumeras sanxit. Carrucas Romae et raedas senatoribus omnibus ut argentatas haberent, permisit, interesse Romanae dignitatis putans, ut his tantae urbis senatores uterentur.*

> He enacted laws without number. He permitted every senator to use a carriage in the city and to have a coach ornamented with silver, thinking that it enhanced the dignity of Rome that these should be used by the senators of so great a city.
> (*Historia Augusta, Severus Alexander* 43.1)

Such wagons would have been used by the senators mainly to impress the people. For speed, it made no difference whether one went on foot or by wagon in Rome. Neither did the emperor Aurelian forbid the use of beautiful wagons of bronze and ivory:

> *Dedit praeterea potestatem, ut argentatas privati carruchas haberent, cum antea aerata et eburata vehicula fuissent.*

> He furthermore granted permission to commoners to have coaches adorned with silver, whereas they had previously had only carriages ornamented with bronze or ivory.
> (*Historia Augusta, Aurelianus* 46.3)

It remained uncommon to drive by wagon in Rome by day, apart from the aforementioned exceptions. That the *Lex Julia Municipalis* was obviously in force in Antioch is demonstrated by the fact that its inhabitants did not like wagons as part of the street scene. Rubble was also disposed of by means of animals.[892] Aurelian, wounded when he approached the city, descended from his wagon and entered the city on horseback.[893] After Aurelian, in the fourth century, beautiful wagons became more usual for upper-class officials as a means of transport.[894]

As mentioned,[895] the aediles were responsible for street maintenance, but who had to tackle traffic offenders is not clear. The *vigiles* were the best qualified, but their task was in fact the maintenance of public safety and fire fighting, similar to our contemporary police and fire brigade. In Rome there was a contingent of some 3000 men who attempted both day and night to maintain order, run by the *praefectus vigilum*. Traffic abuses were also settled by them. The *vigiles* had their barracks at Porta Ratumena. Serious traffic abuses were dealt with at the *praefectus urbi*, at the Forum Augusti.[896]

Summary and conclusion

Initially, traffic did not cause chaotic situations. However, as a result of the growing population, the road network and logistics forced Roman governments to pass laws to maintain control of the traffic flow. The most well-known law is the *Lex Julia Municipalis*, introduced around the middle of the first century BC. The law resulted in a drastic reduction of wheeled traffic by day, but with many exceptions: in particular building traffic and disposal of rubble, excrement and waste still caused much traffic in the city. Claudius commanded that within cities (in Italy) only pedestrian traffic was permitted by day.

It must have been almost impossible to maintain these laws; later emperors had to ratify them repeatedly, particularly the decree of Claudius. Until Late Antiquity, it was busy in the streets and the problem was in fact not solved merely by legislation, any more than it is nowadays.

Traffic circulation

Introduction

Apart from legislation, a local government still has another means of regulating traffic: by introducing physical obstacles, forcing road-users to take stipulated routes. This occurred exclusively in cities; interurban traffic between two cities usually took place along one road, without alternative routes.[897] There was, generally, no need to impose restrictions on interurban traffic. In Classical Greece, in the neighbourhood of Eleusis, a traffic restriction was installed by means of a narrow bridge (5 Greek feet, 1.55 m) to keep the ceremonies accessable to pedestrians.[898]

In an urban context, the situation was different. Hippodamic-planned cities had, by definition, a number of streets running parallel to each other, from which road-users could choose. We can assume that road-users – just like today – preferred the shortest route, but a city government could even choose to block the route for through-traffic, so that it was forced to take another route instead of the shortest and most efficient one.

In this chapter, traffic circulation of an old city (Pompeii) and a later founded *colonia* (Xanten) will be discussed.

Traffic circulation in Pompeii

Pompeii started as a settlement with an irregular street pattern. This street pattern was maintained during the whole history of Pompeii in the old city centre, the districts VII and VIII, where the *forum* is also situated. A first extension took place in the fifth century BC, where district VI was planned with a regular, although not rectangular, street system. On the *forum*, a former *cardo*, Via di Mercurio, was connected, leaving the city through a gate where Torre XI (Torre di Mercurio) was built. At the second extension stage another new *cardo* was built, Via Stabiana, from Porta di Stabia to Porta del Vesuvio; the *decumanus* was extended to Porta di Sarno. The *forum* was therefore no longer the point of intersection of *cardo* and *decumanus*, but remained in the old city. On the east side of the *cardo*, the districts IV and IX were planned.

In the House of Menander, a wagon has been found with an axe width from wheel to wheel of 1.42 m;[899] the cross-section of the wheels is 3 cm, and from hub to hub the distance is 1.79 m. Other wagons found in the neighbourhood have a small deviation, but they must have been able to pass the stepping stones without much effort.[900] For a wagon to be able to drive through a street, not only did the street have to be wide enough for the wheels, but the edges of the sidewalks also had to be no higher than the hubs.[901]

Adopting this information, we have to deduce that only a few streets in Pompeii were wide enough for two-way traffic, a breadth of about 3.60 m: Via dell'Abbondanza, Via Stabiana and Via di Nola. Two-way traffic was also possible along some (short) street sections and wagons were able to pass each other, for instance between Porta di Nocera and Via di Castricio. In some places, for example Via di Nola, pavements were narrowed for the protection of sidewalk edges and fountains;[902] this situation also occurs in Via dell'Abbondanza (Fig. 4.1). In a busy street, for instance Via Stabiana, stepping stones were worn by the traffic (Fig. 4.2). The other streets either were only arranged for one-way traffic, or were completely blocked for traffic. The latter occurred by the putting up of stone barricades. Such barricades are found amongst others between *insulae* 10 and 12 district VI and between *insulae* 4 and 3 district IX, where streets were blocked on both sides. Wheel ruts in the road surface show that the obstacles were placed after traffic had formerly driven over the area.[903]

In a city where there is mainly one-way traffic, one can expect a circulation which follows stipulated routes. Passing is therefore impossible. According to Mau, a road-user could wait until the street was free, but once in the street, nevertheless, one could meet oncoming traffic; whereupon one of them had to reverse.[904] A cumbersome matter, especially when behind them new road-users had arrived on the scene.

Figure 4.1 Pompeii, Via dell'Abbondanza, with stepping stones and protected fountain. Photo: C.R. van Tilburg.

One can therefore confirm that certain routes were in use. Another indication for this is formed by the wheel ruts which have been found at the obtuse angles of the *insulae* on both sides of Via Stabiana; wheel ruts at sharp angles are considerably scarcer. Tsujimura states that – if it was prohibited or in any case difficult and unusual to take a sharp bend – a wagon coming from Porta di Stabia with the destination Via di Nola or Via dell'Abbondanza had to make a detour, turning through the old city centre and then, coming from the north, passing the obtuse angle of *insula* 1 district V or *insula* 1 district IX.[905] This detour also passed the concentrated workshop area in the districts VII and VIII and the supply routes will have run as shown in Figure 4.3.

Combined with the fact that there were many barricades, wheeled traffic was forced to take fixed routes. It is, however, not clear which driving-direction

Figure 4.2 Pompeii, worn stepping stones in Via Stabiana. Photo: C.R. van Tilburg.

was in use. Traffic signs comparable with our traffic signs today have not yet been found in Pompeii. They were completely absent or they were destroyed during the eruption of Vesuvius. If there were no traffic signs and fixed directions, Mau's proposition is not completely wrong when he assumes that wagon-assistants preceded the wagon. When a street section with a length of an *insula* was free, a preceding assistant (possibly with a whip) could pass this part of the street rapidly and give the 'safety signal' to the driver coming behind. The preceding assistant could also request oncoming traffic to wait whilst his wagon was passing through part of the street.[906] Crossing and overtaking were possible at every side-street corner. Because most wheeled traffic took place by night – subjected to the *Lex Julia Municipalis* – the torch was the most suitable instrument for a signal. Finding the route was facilitated by the presence of tiles, functioning as street name plates.[907]

As well as the aforementioned blocks and narrowings in the centre of the streets, in the case of road junctions stones were also erected to make driving in the middle of the road impossible. Such stones are found in three areas: in Vicolo di Tesmo *insulae* 1, 2 and 7 district IX, in Via del Labirinto/Via della Fortuna *insulae* 12 and 13 district VI, and at the north side of the *forum*.[908] The *forum* itself was not accessible at all for wheeled traffic (Fig. 4.4); all entrances were blocked[909] and the place was completely surrounded by colonnades. Building traffic – of which there must have been

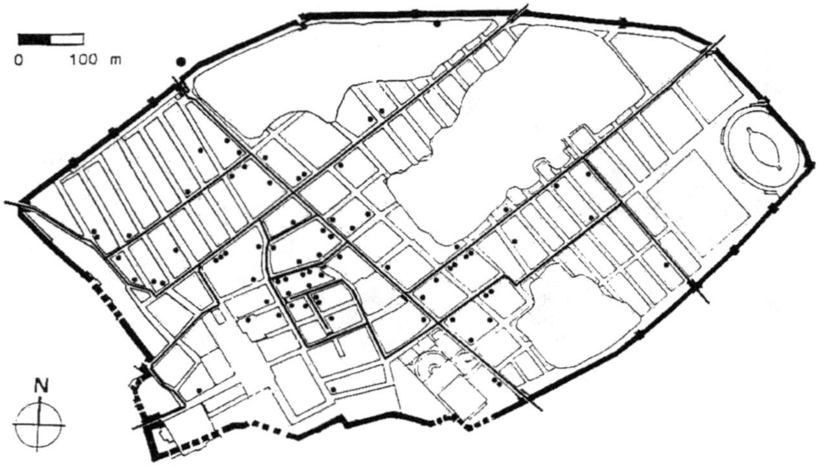

Figure 4.3 Pompeii, possible routes for supplying workshops. Laurence 1994/1996, 66. London: Routledge/C.R. van Tilburg.

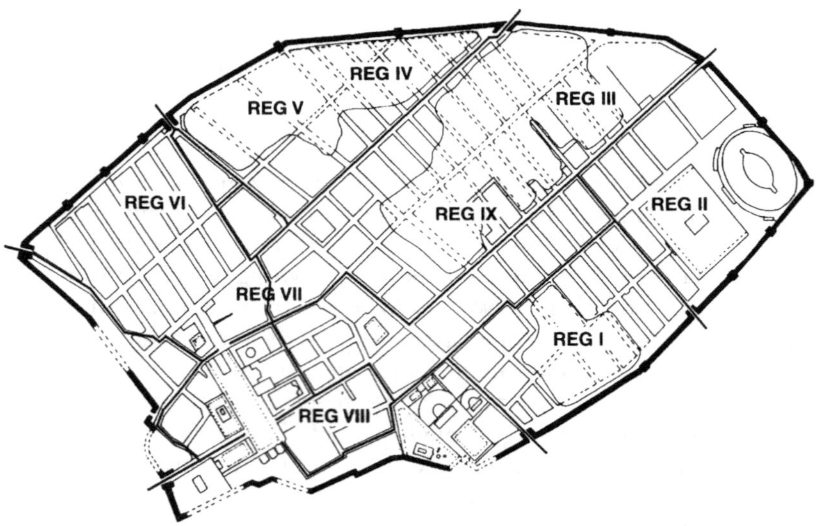

Figure 4.4 Pompeii, possible routes to the *forum*. Picture: C.R. van Tilburg.

a lot after the earthquake of 62! – could only reach the *forum* when the blocks were removed or slopes were used. On the square itself, building traffic would have been rather scarce, but the cul-de-sacs bordered on the colonnades and were accessible to wheeled traffic.

It is striking that in the west part of the city (the districts VI, VII and VIII) wheel ruts are considerably shallower than in the east part, east of Via

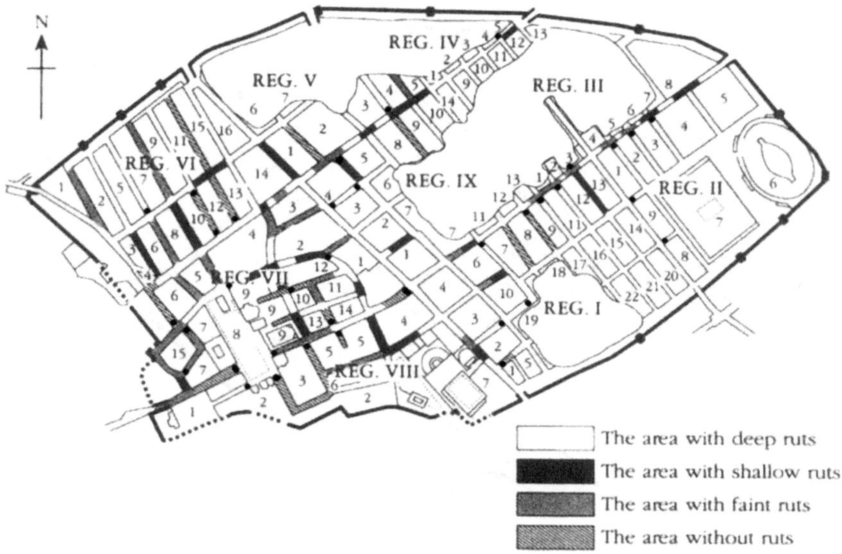

Figure 4.5 Pompeii, blocks and wheel ruts. Wallace-Hadrill 49. London: UCL.

Stabiana. In other words: to the east of Via Stabiana there are more streets with deep ruts than west of it. East of Via Stabiana and on Via Stabiana itself, more wheel ruts are also found at obtuse corners.

According to Wallace-Hadrill, the difference in depth of the wheel ruts between the west and the east part of the city is determined by the presence of public buildings and taverns. On the west side, most of the public buildings are situated – except the amphitheatre and the Terme Centrali – and the city government (the aediles) chose to establish the taverns elsewhere. Up to now, they have mostly been found at Porta di Ercolano, Porta di Stabia, Via Consolare, Via dell'Abbondanza (east of the intersection with the Via Stabiana), Via di Nola, Via del Vesuvio and Via Stabiana.[910] In district VII, on the other hand, there was a concentration of workshops.[911] Here, however, the wheel ruts are not so deep, which could indicate that traffic was less dependent on wheeled traffic or was less intensive (Fig. 4.5).

Why were so many streets blocked? Not merely to hinder and to make the use of certain routes impossible. This argument is valid for barricades around the *forum*; the *forum* was, like the imperial *fora* in Rome, completely closed for wheeled traffic (Fig. 4.6). The barricade in Via dell'Abbondanza (Fig. 4.7), on the west part of the intersection with Via Stabiana, was probably set up for traffic safety reasons. This point, the intersection of *cardo* and *decumanus* in Pompeii, must have been one of the busiest places of the city. This, in combination with the steep slope of Via dell'Abbondanza, motivated the

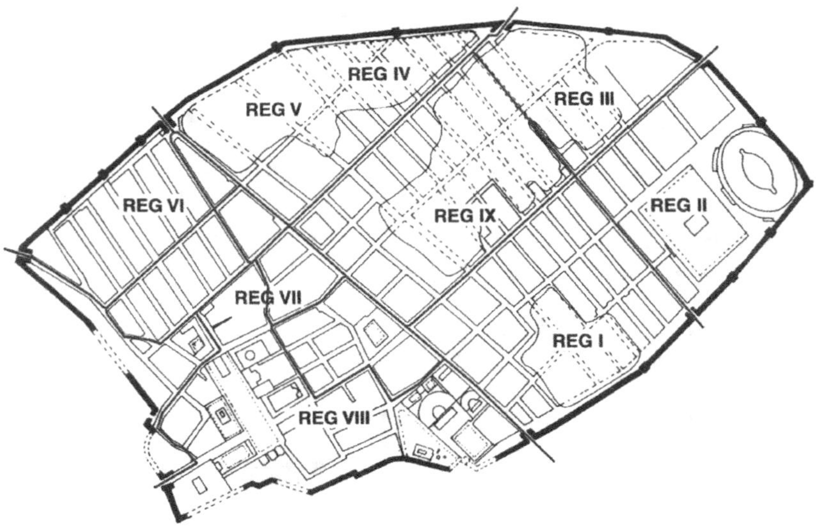

Figure 4.6 Pompeii, possible through-routes. Picture: C.R. van Tilburg.

local government to erect the barricade. But the barricade resulted in radical consequences for traffic circulation.

Apart from these barricades, blocking wheeled traffic completely, obstacles are also found hindering traffic, although not blocking it completely. They are found in districts VII and VIII. The fact that these obstacles could be passed by wheeled traffic is proven by the presence of wheel ruts.[912]

The side streets connecting Via dell'Abbondanza in districts I and II were nearly all blocked and provided with grilles which could be closed; these were placed by both city government and private individuals. In districts III, IV and IX these grilles are absent,[913] which gives the impression that inhabitants had some influence in creating traffic-hindering obstacles. It is exactly in the districts II, III and IV that the majority of the streets have deep wheel ruts caused by traffic supplying taverns.[914]

There were also other reasons for creating barricades, to do with water management. Building an underground sewer for the discharge of rainwater in a city like Pompeii, built on volcanic soil (the pavement is basalt), was almost impossible. In the case of showers rainwater had to be removed as soon as possible; the many stepping stones (a characteristic of Pompeii; in Herculaneum and Paestum they are completely absent)[915] were in the first instance to make it possible for pedestrians to cross the street without getting wet and muddy feet. That thereby the difference in level between curb and street was raised was an agreeable added detail.

The reason for barricades in the side streets of Via dell'Abbondanza could

Figure 4.7 Barricade in Via dell'Abbondanza, intersection with Via Stabiana. Photo: C.R. van Tilburg.

have been to prevent rainwater from flowing into these side streets. At one point, at *insulae* 8 and 9, district I, the barricade was accommodated with a drainage channel where rainwater could flow through, but not the traffic. So the interests of traffic were less important than the interests of water management.[916]

Traffic circulation in Xanten

In Pompeii, traffic could function when road-users took fixed routes. One must assume that this statement is valid for all cities which had an organic growth, especially when streets were narrow. One-way traffic was not disadvantageous to the logistics if it flowed well. But in Antiquity there must also have been the aspiration to build streets wide enough to enable wagons to pass or overtake each other without problems. Such a system had the advantage of limiting traffic congestion. So the Romans, in the case of planned cities, built streets wide enough for two-way traffic. In Bavay, streets have been found with a width varying from 4.50 m to 6.20 m,[917] wide enough for two-way traffic. Other cities in northern Europe have similar streets.

As mentioned in chapter 1 (p. 31), Xanten was accommodated with streets 10–12 m wide, flanked by colonnades with a width of about 4 m; in this respect, Xanten was in the lead as far as wide streets were concerned.[918] One would expect that there were no restrictions; not only was two-way traffic possible everywhere, but neither did parking in the streets cause many problems. So we can assume that Xanten – as far as wheeled traffic was concerned – was an ideal city, without one-way traffic, narrow lanes or barricades. Because the streets were surfaced with gravel and rainwater therefore disappeared rapidly into the ground or left the city via underground sewers, there was no need for stepping stones, typical in Pompeii. The Hippodamic street system caused an unimpeded situation with unobstructed corners nearly everywhere. There would have been no urgent need for fixed routes.

Nevertheless, there is an indication that there was one particular route. On the east side, bordering the harbour quay running parallel with the city wall, there were three harbour gates: Hafentor, Molentor and Kleine Hafentor. These gates were too small for two-way traffic but were intended for one-way traffic. They were probably the only points where there was one-way traffic. Goods transported by ship along the Rhine entered the city through one gate, were delivered, and the wagons – either empty or with a return load – left the city through another gate, arriving back in the harbour area.

The most southerly harbour gate, Kleine Hafentor, had, as mentioned, one single passage. The gate was small and it caused a lot of hindrance to the drivers of wagons and animals in passing the gate and, further on, leaving the gate, turning the bend. Evidence is shown by the heavily worn cornerstone at the base of the north side of the city-side of the gate (Figs 4.8 and 4.9). This indicates that busy traffic used this gate.

Looking at the plan of Xanten, the river side was sloped in relation to the *cardo* and, inevitably, caused sharp and obtuse angles in the *insulae*. When there was a similar situation in Pompeii, namely that wheeled traffic preferred to turn obtuse corners, the most logical route for supplying the *forum* would have been as shown in Figure 4.10. In the Hafentempel area the *pomerium* was blocked, so wagons could not turn the obtuse corner here. They were forced to go straight on, leaving the city. Traffic coming from Molentor could – in respect of the infrastructure – go straight on without problems, in entering and leaving the city, but this gate was probably also part of a one-way traffic route. Kleine Hafentor was too narrow for some loads.

Summary and conclusion

There was a definite flow of traffic in both old and latterly built Roman cities, which was used intensively. Older cities like Pompeii were mainly set up as a one-way traffic area with narrow streets. Two-way streets were limited. The consequence was that traffic had to select fixed routes. The local government also enforced routes by erecting barricades, where wheeled traffic had earlier been possible. Furthermore routes were stipulated by incorporating obtuse angles as much as possible into these routes. Turning sharp bends was difficult, and perhaps in many cases also forbidden. In spite of these

Figure 4.8 Xanten, Kleine Hafentor with the worn cornerstone on the left. Photo: C.R. van Tilburg.

restrictive measures, there was a lot of wheeled traffic in Pompeii, as proven by the sometimes deep wheel ruts in basalt, a very hard type of stone.

The situation in these cities must have been comparable with historical Dutch and English city centres nowadays, where a one-way system is also frequently used. A big difference however is that in the Dutch and English

Figure 4.9 Xanten, worn cornerstone. Photo: C.R. van Tilburg.

city centres the prescribed routes are indicated by signs. Indications of the use of traffic signs in a city like Pompeii have, as far as we know, not been found.

In later, Hippodamic cities there were also fixed routes. In Xanten a cornerstone of Kleine Hafentor has been found that has been worn away by busy traffic use. Here too traffic preferred to use an obtuse bend. We can therefore state that obtuse bends were an important factor in prescribing a route.

Missed chances and political lack of interest

Introduction

In the previous sections it is stated that there were traffic restrictions and fixed routes in both old and later planned cities. The authorities tried to keep the amount of traffic under control by means of legislation and the erection of barricades in streets. In fact this was only tackling the symptoms.

In this final section we will discuss whether, and if so to what extent, (local) governments tried to tackle the traffic problems. Apart from legislation and control of existing traffic flow, traffic congestion can also be tackled by infrastructure improvement. From the nineteenth century to the present day sticking points have been tackled by creating space for traffic. Canals have been filled in, gates demolished, streets broadened and houses knocked

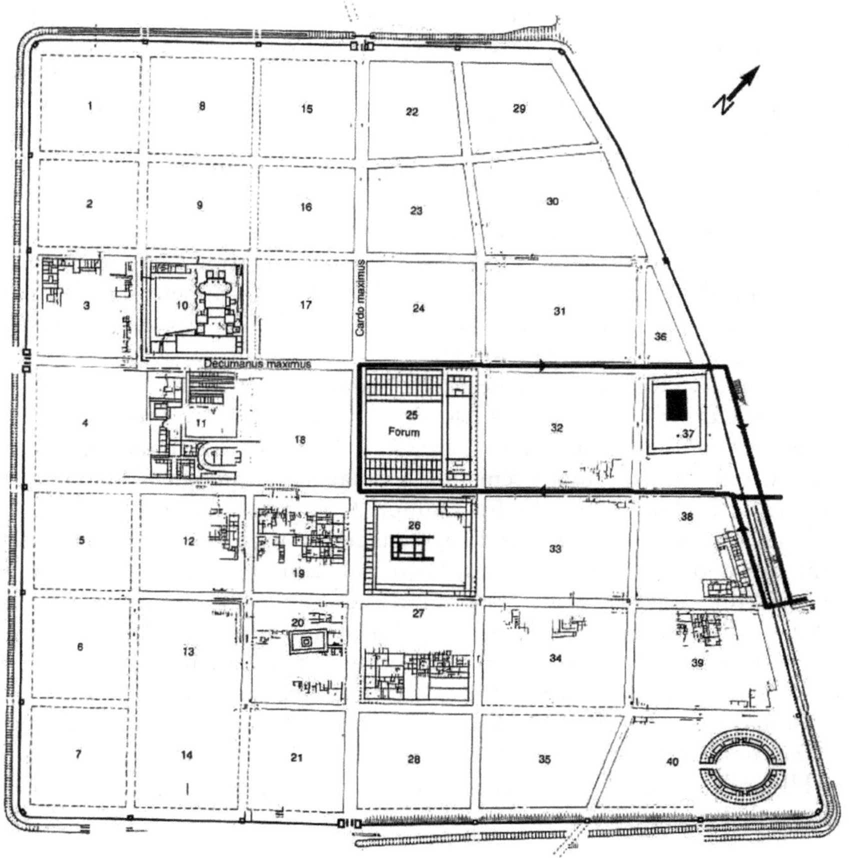

Figure 4.10 Xanten, possible routes from the harbour to the *forum*. Heimberg and Rieche 7. H. Stelter, APX/C.R. van Tilburg.

down. In our time we are still trying to find alternatives for the increasing car traffic. Have examples also been found of such measures in the Roman Empire?

The existing situation was maintained

One can generally state that through the course of the history of a city the street system does not radically change. Historical European cities (like Leiden, Haarlem or Amsterdam in the Netherlands and Canterbury and Salisbury in Britain) preserved the same street patterns from their beginnings up to the present time – with the exception of certain details.

This situation becomes quite different when a city is devastated, by the violence of war, fire or a natural disaster. A city government can then make a choice between two possibilities: maintaining the former situation – wholly or partially – or making a completely new start. Whichever possibility is chosen depends on a number of factors: the degree of devastation, the desire to preserve former structures (including the street plan), and the requirements of the enemy in the case of war, after the sack of the city.

The building of an entire new city with little or no reference to the former situation was and remains an exception.[919] When the choice has been made to maintain the existing city, the centre where the government was situated was in any case maintained where possible. To keep this centre accessible, the approach roads had to be maintained.

In Antiquity cities were also destroyed. Some examples are Troy, Athens, Carthage, Corinth, Bavay and Cologne. Carthage and Corinth, both destroyed by the Romans in 146 BC, are examples of total devastation. But in most cases the street pattern remained unchanged, although in some cases a period of decline followed – for example Cologne, devastated in 355 by the Franks. Another example is Tongeren. In the years 68–70 the city was burnt down by the Batavians, but with the rebuilding the street pattern remained the same. Only after the fall of the Roman Empire did the street pattern radically change.[920] Also Bavay, burnt down between 180 and 190, preserved its former street pattern. Obviously, the street patterns were satisfactory; the opportunity to change and to improve them was not taken.[921] In Xanten, some streets were maintained dating from the time before the *colonia*; these streets are still visible on the plan as streets which do not follow the Hippodamic lines of the chessboard pattern exactly and do not have right-angled corners.[922]

An attempt at city renewal: Neropolis

In the previous chapters it has been mentioned that Rome suffered from its problems with city planning and traffic congestion. The causes are known: the chaotic street pattern caused by the devastation of the Gauls in 390 BC, the presence of hills and a winding river, high land prices stimulating the building of tall apartment buildings and narrow lanes, and the uncontrolled growth of the city since the middle of the second century BC.

In the year 64 AD, Rome was struck by a devastating fire, destroying the majority of the city. The question of whether Nero himself was responsible for this disaster will not be discussed here, but Tacitus and Suetonius describe how Nero was not only unperturbed by the fire, but also responsible for it. He immediately took the chance of putting his ideas on city planning into practice, including the creation of a megalomaniacal palace, the Domus Aurea, the 'Golden House'. Here follows the text and translation of Suetonius:

Sed nec populo aut moenibus patriae pepercit. Dicente quodam in sermone communi: ''Εμοῦ θανόντος γαῖα μειχθήτω πυρί', 'Immo', inquit, 'ἐμοῦ ζῶντος', planeque ita fecit. Nam quasi offensus deformitate veterum aedificiorum et angustiis flexurisque vicorum, incendit urbem tam palam, ut plerique consulares cubicularios eius cum stuppa taedaque in praediis suis deprehensos non attigerint, et quaedam horrea circum domum Auream, quorum spatium maxime desiderabat, ut bellicis machinis labefacta atque inflammata sint, quod saxeo muro constructa erant. Per sex dies septemque noctes ea clade saevitum est ad monumentorum bustorumque deversoria plebe compulsa. Tunc praeter immensum numerum insularum domus priscorum ducum arserunt hostilibus adhuc spoliis adornatae deorumque aedes ab regibus ac deinde Punicis et Gallicis bellis votae dedicataeque, et quidquid visendum atque memorabile ex antiquitate duraverat. Hoc incendium e turre Maecenatiana prospectans laetusque 'flammae', ut aiebat, 'pulchritudine' Halosin Ilii in illo suo scaenico habitu decantavit. Ac ne non hinc quoque quantum posse praedae et manubiarum invaderet, pollicitus cadaverum et ruderum gratuitam egestionem nemini ad reliquias rerum suarum adire permisit; conlationibusque non receptis modo verum et efflagitatis provincias privatorumque census prope exhausit.

But he showed no greater mercy to the people or the walls of his capital. When someone in a general conversation said: 'When I am dead, be earth consumed by fire,' he rejoined 'Nay, rather while I live,' and his action was wholly in accord. For under cover of displeasure at the ugliness of the old buildings and the narrow, crooked streets, he set fire to the city so openly that several ex-consuls did not venture to lay hands on his chamberlains although they caught them on their estates with tow and firebrands, while some granaries near the Domus Aurea, whose room he particularly desired, were demolished by engines of war and then set on fire, because their walls were of stone. For six days and seven nights destruction raged, while the people were driven for shelter to monuments and tombs. At that time, besides an immense number of dwellings, the houses of leaders of old were burned, still adorned with trophies of victory, and the temples of the gods vowed and dedicated by the kings and later in the Punic and Gallic wars, and whatever else interesting and noteworthy had survived from antiquity. Viewing the conflagration from the tower of Maecenas and exulting, as he said, in 'the beauty of the flames,' he sang the whole of the 'Sack of Ilium,' in his regular stage costume. Forthermore, to gain from this calamity too all the spoil and booty possible, while promising the removal of the debris and dead bodies free of cost he allowed no one to approach the ruins of his own property; and from the contributions which he not only received, but even demanded, he nearly bankrupted the provinces and exhausted the resources of individuals.'

(Suet. *Nero* 38.1–3)

A more extensive and impressive description of the fire is given by Tacitus:

> [38] *Sequitur clades, forte an dolo principis incertum (nam utrumque auctores prodidere), sed omnibus, quae huic urbi per violentiam ignium acciderunt, gravior atque atrocior. Initium in ea parte circi ortum, quae Palatino Caelioque montibus contigua est, ubi per tabernas, quibus id mercimonium inerat, quo flamma alitur, simul coeptus ignis et statim validus ac vento citus longitudinem circi corripuit. neque enim domus munimentis saeptae vel templa muris cincta aut quid aliud morae interiacebat. Impetu pervagatum incendium plana primum, deinde in edita adsurgens et rursus inferiora populando, anteiit remedia velocitate mali et obnoxia urbe artis itineribus hucque et illuc flexis atque enormibus vicis, qualis vetus Roma fuit. ad hoc lamenta paventium feminarum, fessa aut rudis pueritiae aetas, quique sibi quique aliis consulebant, dum trahunt invalidos aut opperiuntur, pars mora, pars festinans, cuncta impediebant. Et saepe, dum in tergum respectant, lateribus aut fronte circumveniebantur, vel si in proxima evaserant, illis quoque igni correptis, etiam quae longinqua crediderant in eodem casu reperiebant. Postremo, quid vitarent quid peterent ambigui, complere vias, sterni per agros; quidam amissis omnibus fortunis, diurni quoque victus, alii caritate suorum, quos eripere nequiverant, quamvis patente effugio interiere. Nec quisquam defendere audebat, crebris multorum minis restinguere prohibentium, et quia alii palam facies iaciebant atque esse sibi auctorem vociferabantur, sive ut raptus licentius exercerent seu iussu.*
>
> [39] *Eo in tempore Nero Antii agens non ante in urbem regressus est, quam domui eius, qua Palatium et Maecenatis hortos continuaverat, ignis propinquaret. Neque tamen sisti potuit, quin et Palatium et domus et cuncta circum haurirentur. sed solacium populo exturbato ac profugo campum Martis ac monumenta Agrippae, hortos quin etiam suos patefecit et subitaria aedificia exstruxit, quae multitudinem inopem acciperent; subvectaque utensilia ab Ostia et propinquis municipiis, pretiumque frumenti minutum usque ad ternos nummos. Quae quamquam popularia in inritum cadebant, quia pervaserat rumor ipso tempore flagrantis urbis inisse eum domesticam scaenam et cecinisse Troianum excidium, praesentia mala vetustis cladibus adsimulantem.*
>
> [40] *Sexto demum die apud imas Esquilias finis incendio factus,*

prorutis per immensum aedificiis, ut continuae violentiae campus et velut vacuum caelum occurreret. necdum positus metus aut redierat plebi spes: rursum grassatus ignis patulis magis urbis locis, eoque strages hominum minor: delubra deum et porticus amoenitati dicatae latius procidere. Plusque infamiae id incendium habuit, quia praediis Tigellini Aemilianis proruperat; videbaturque Nero condendae urbis novae et cognomento suo appellandae gloriam quaerere. Quippe in regiones quattuordecim Romam dividitur, quarum quattuor integrae manebant, tres solo tenus deiectae: septem reliquis pauca tectorum vestigia supererant, lacera et semusta.

[41] *Domuum et insularum et templorum, quae amissa sunt, numerum inire haud promptum fuerit; sed vetustissima religione, quod Servius Tullius Lunae, et magna ara fanumque, quae praesenti Herculi Arcas Evander sacraverat, aedesque Statoris Iovis vota Romulo Numaeque regia et delubrum Vestae cum Penatibus populi Romani exusta; iam opes tot victoriis quaesitae et Graecarum artium decora, exim monumenta ingeniorum antiqua et incorrupta, ut quamvis in tanta resurgentis urbis pulchritudine multa seniores meminerint, quae reparari nequibant. Fuere qui adnotarent XIIII Kal. Sextiles principium incendii huius ortum, quo et Senones captam urbem inflammaverint. Alii eo usque cura progressi sunt, ut totidem annos mensesque et dies inter utraque incendia numerent.*

[42] *Ceterum Nero usus est patriae ruinis exstruxitque domum, in qua haud perinde gemmae et aurum miraculo essent, solita pridem et luxu vulgata, quam arva et stagna et in modum solitudinem hinc silvae, inde aperta spatia et prospectus, magistris et machinatoribus Severo et Celere, quibus ingenium et audacia erat etiam, quae natura denegavisset, per artem temptare et viribus principis inludere. . . .*

[43] *Ceterum urbis quae domui supererant non, ut post Gallica incendia, nulla distinctione nec passim erecta, sed dimensis vicorum ordinibus et latis viarum spatiis cohibitaque aedificiorum altitudine ac patefactis areis additisque porticibus, quae frontem insularum protegerent. Eas porticus Nero sua pecunia exstructurum purgatasque areas dominis traditurum pollicitus est. Addidit praemia pro cuiusque ordine et rei familiaris copiis, finivitque tempus, intra quod effectis domibus aut insulis apiscerentur. Ruderi accipiendo Ostiensis paludes destinabat, utique naves, quae frumentum Tiberi subvectassent, onustae rudere decurrerent, aedificiaque ipsa certa sui parte sine trabibus saxo Gabino Albanove solidarentur, quod is lapis ignibus impervius est; iam aqua privatorum licentia intercepta quo largior et pluribus locis in publicum flueret, custodes adessent; et subsidia reprimendis ignibus in propatulo quisque haberet; nec communione parietum, sed propriis quaeque muris ambirentur. Ea ex utilitate accepta decorem quoque novae urbi attulere. Erant tamen qui crederent, veterem illam formam salubritati*

magis conduxisse, quoniam angustiae itinerum et altitudo tectorum non perinde solis vapore perrumperentur: at nunc patulam latitudinem et nulla umbra defensam graviore aestu ardescere.

What happened next was a disaster and it is not known whether this was an accident or in fact arson committed by the Emperor out of maliciousness – there are varied opinions – but whatever it was, this was the worst fire which had ever taken place in the city. It started in the shops which were full of inflammable goods, near where the Circus adjoins the Palatine and Caelian Hills, and of course it didn't take long before it spread along the full length of the Circus, fanned by the wind. The houses and temples in the area were not surrounded by stone walls or any other type of protection, so the initial progress of the fire went unchecked. The first flames made short work of the low-lying districts in the immediate vicinity, climbing to higher areas and then descending again to destroy more districts situated at ground level. Any attempts at firefighting were pointless; it surged ahead, travelling faster and faster, continually gaining ground, its progress made easy by the twisting narrow lanes and streets which were originally a characteristic of old Rome. Screaming and terrified women, the elderly or invalid, children too young to know what was going on, people running to save either themselves or others, people dragging along those who were too infirm or having to stop and wait for them, all aggravated the general confusion and terror. When one looked backwards, one was attacked from the side or the front; when one finally managed to escape to what looked like a safer place away from the flames, that also turned out to be a mistake. Finally, in desperation and no longer knowing what to do, people thronged into the streets or threw themselves down into the fields; those who had lost their whole livelihoods including all their food decided to give up the fight, even though an escape-route was actually still open to them. Others took the same tragic decision who had seen their loved ones perish or who had been unable to save them. Strangely, no-one made any attempts to fight the fire because there were continual threats from certain members of the community who actually forbade it to be extinguished. Others were also visibly throwing firebrands in order to aggravate the flames, yelling that they had permission to do so. This could have been to facilitate looting, or it could also have been that someone was indeed giving orders to let the blaze continue.

Nero was staying in Antium at the time. He didn't return to Rome until the fire was nearing the house which he had built to join the Palatine with the Gardens of Maecenas. However, it proved impossible to stop the fire from swallowing up both the Palatine, the house, and everything in the near vicinity. In a gesture to the masses of homeless and those on the run, he opened the Campus Martius, the Agrippa buildings, as well as his own gardens, and erected shelters as temporary housing. Food supplies were brought up from Ostia and the neighbouring towns and villages and the price of grain was lowered to three sesterces. Although these were popular gestures, they had no effect because word then went round that when the fire started, Nero had been seen climbing his platform, singing of the destruction of Troy, and equating the disasters of the present with the trials and tribulations of the past.

Finally, after raging for six days, the fire was extinguished at the foot of the Esquiline Hill. This was done by demolishing buildings over a vast area so that there was nothing left to burn. However, the people were terrified anew when it re-started in another part of the city which was less congested. The death-toll was albeit smaller, but the destruction of temples and other buildings used for pleasure purposes was far greater. This second fire actually triggered a rumour and by far the greater scandal of the two: the emerging theory that because it had broken out on the Aemilian property of Tigellinus it suggested that Nero could possibly be attempting to found a new city to be inaugurated in his own name, and this would appear to support the arson theory. Rome is divided into fourteen districts; four remained intact, three were razed to the ground, and in the remaining seven little remained except for a few dilapidated and half burnt-out houses.

It would be impossible to give anywhere near an estimate of the private homes, apartment blocks, temples and suchlike, which were lost. The flames destroyed the temple dedicated to Luna by Servius Tullius, the high altar and shrine of the Arcadian Evander to the Contemporary Hercules, the shrine of Jupiter the Steadfast consecrated by Romulus, the Palace of Numa, Vesta's Sanctuary with the deities of the Roman people; all these were burnt in what seemed an almost divine sanctification. Trophies which had been won in battle, glorious Greek art, examples of literary genius; all were destroyed so that in the beautiful new city which then arose to replace it, there were many members of the older generation who could remember places and things which were gone and impossible to replace. It was particularly significant to some that the fire had started on the 19th July, the anniversary of the capture and burning of Rome by the Senones: another theory was that the period of time between the two fires could be divided into equal numbers of years, months and days.

Nero, however, immediately went to work and put the destruction to good use by building a new palace which consisted not just of gold or precious stones which the people were already quite accustomed to, but of parks, lakes and woods, interspersed with open landscape. The architects and engineers commissioned to do the work were Severus and Celer. Clever and inventive, they pitted their wits against nature, going way over the Emperor's budget in the process. They embarked upon the task of sinking a canal from Lake Avernus to the mouth of the Tiber, working along a lonely stretch of shoreline or by boring through hills, crossing a barren, desert area, with the only means of water-supply coming from the Pomptine Marshes. The excruciating labour involved in excavating this dry area was pointless and could have had no worthwhile results. However, Nero, with his love of the impossible, tried to tunnel through the hills near the Avernus, and evidence of this futile attempt can still be seen today.

Meanwhile in the capital any districts not encompassed by Nero's huge new palace were rebuilt. This time (in contrast to the Gallic Fire, when everything was rebuilt indiscriminately and in piecemeal fashion) long, wide streets appeared, with a restriction on the height of the buildings, large open spaces, with colonnades along the fronts of the buildings as added protection. Nero promised the people that the colonnades would be erected at his own expense, and all building-sites would be handed back to the landlords, cleared of wreckage. He also offered a system of rewards, in proportion to the class and wealth of claimants, by fixing a term for completion of house-building, upon which the reward was paid out. The debris from the fire was dumped on the Ostian Marshes, and he gave orders that all ships coming up the Tiber carrying grain, should return downstream laden with debris. All buildings were to be solid, untimbered, and constructed of Gabine or Alban stone which was fireproof. From now on the municipal water-supply, which hitherto could sometimes illegally be intercepted, was to be available to the public in greater quantities and at more points; fire-hydrants were to be kept by everyone out in the open; all buildings had to be detached with no party-walls. All these reforms were welcomed by the public, as well as improving the look of the new city. Obviously there were those who felt that the old situation was better, when the higher buildings and narrower streets offered more protection against the sun. Under the new situation, the wider expanses with no means of shade could be sweltering hot.

(Tac. *Ann.* 15.38–43, tr. J. van Klaveren)

These two descriptions of the fire and the rebuilding of Rome therefore mainly agree. The fire caused many deaths; it broke out at the Circus Maximus; there was resistance against extinguishing it; the majority of

Rome had burnt down; Nero used the released ground to build a new palace with huge gardens and parks and he spent a lot of attention on the rebuilding of the city. As a matter of fact, the city had to carry his name: Neropolis.[923]

New Rome – or Neropolis – was built as a splendidly planned city, with avenues, lower apartment buildings, more fountains, more parks and colonnades. A new palace was built, the Domus Aurea. Suetonius gives the following description:

> *Non in alia re tamen damnosior quam in aedificando domum a Palatio Esquilias usque fecit, quam primo transitoriam, mox incendio absumptam restitutamque auream nominavit. De cuius spatio atque cultu suffecerit haec rettulisse. Vestibulum eius fuit, in quo colossus CXX pedum staret ipsius effigie; tanta laxitas, ut porticus triplices miliarias haberet; item stagnum maris instar, circumsaeptum aedificiis ad urbium speciem; rura insuper arvis atque vinetis et pascuis silvisque varia, cum multitudine omnis generis pecudum ac ferarum. In ceteris partibus cuncta auro lita, distincta gemmis unionumque conchis erant; cenationes laqueatae tabulis eburneis versatilibus, ut flores, fistulatis, ut unguenta desuper spargerentur; praecipua cenationum rotunda, quae perpetuo diebus ac noctibus vice mundi circumageretur; balineae marinis et albulis fluentes aquis. Eius modi domum cum absolutam dedicaret, hactenus comprobavit, ut se diceret quasi hominem tandem habitare coepisse.*
>
> There was nothing however in which he was more ruinously prodigal than in building. He made a palace extending all the way from the Palatine to the Esquiline, which at first he called the House of Passage, but when it was burned shortly after its completion and rebuilt, the Domus Aurea. Its size and splendour will be sufficiently indicated by the following details. Its vestibule was large enough to contain a colossal statue of the emperor a hundred and twenty feet high; and it was so extensive that it had a triple colonnade a mile long. There was a pond too, like a sea, surrounded with buildings to represent cities, besides tracts of country, varied by tilled fields, vineyards, pastures and woods, with great numbers of wild and domestic animals. In the rest of the house all parts were overlaid with gold and adorned with gems and mother-of-pearl. There were dining-rooms with fretted ceilings of ivory, whose panels could turn and shower down flowers and were fitted with pipes for sprinkling the guests with perfumes. The main banquet hall was circular and constantly revolved day and night, like the heavens. He had baths supplied with sea water and sulphur water. When the edifice was

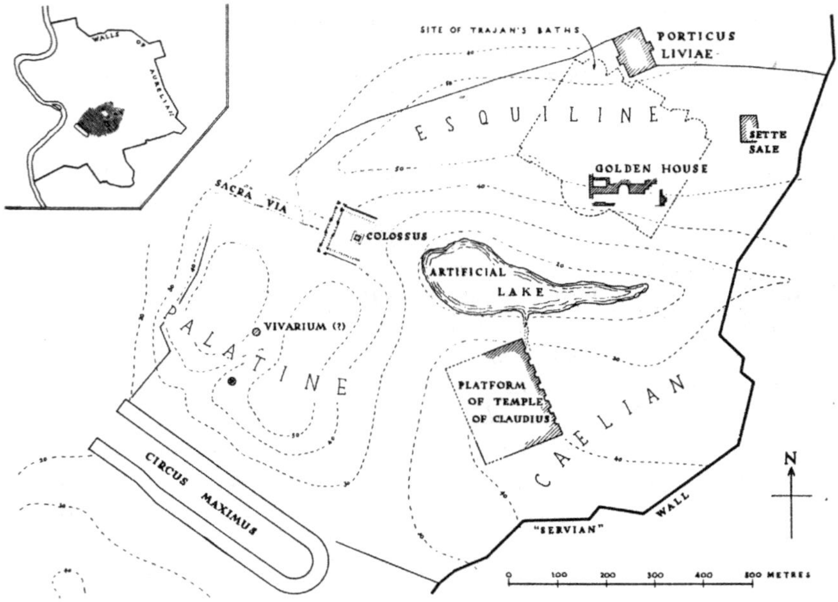

Figure 4.11 After the fire: the Domus Aurea with parks and water. Bergmann 20. Mainz: von Zabern.

> finished in this style and he dedicated it, he deigned to say nothing more in the way of approval than that he was at last beginning to be housed like a human being.
>
> (Suet. *Nero* 31.1–2)

Suetonius must have been exaggerating; the city could not have had a 'pond like a sea' (*stagnum maris instar*). Perhaps not the Domus Aurea itself but the whole area extended from Palatine to Esquiline (Fig. 4.11).[924] The survivors of the fire saw a palace and gardens at the place where their homes once stood, so they did not like the vast and impressive Domus Aurea.[925]

We know of one possible image of Neropolis. In 1998, beneath the Baths of Trajan, to the north-east of the Colosseum, a fresco was found representing a bird's eye view (Fig. 4.12). What we see is a city surrounded by a wall, on the left a theatre and a statue (of Apollo) and on the right a square or rectangular building with a water basin. The wall is furnished with round towers and to the left behind the wall is a bridge. If we accept this is a view of Neropolis, it is not an exact reproduction of the buildings; in other words, the theatres of Rome – of Marcellus, Pompeius and Balbus – are all situated

Figure 4.12 Neropolis, fresco. Van der Meer 70. Leiden: Rijksmuseum van Oudheden.

outside the Servian Wall and not inside. Moreover, the Servian Wall had already disappeared for part of Nero's reign.[926]

So Neropolis was a symbol of Nero's importance. But did Nero have any intention of improving the traffic situation by means of these enormous projects? Neither Tacitus nor Suetonius mentions better traffic flow. Tacitus speaks of 'wide streets' (*latis viarum spatiis*) but it is not certain whether these streets were constructed to sustain more traffic. There was even criticism of these streets: they were too broad, there was no shadow against the bright sun (*Erant tamen qui crederent, veterem illam formam salubritati magis conduxisse, quoniam angustiae itinerum et altitudo tectorum non perinde solis vapore perrumperentur: at nunc patulam latitudinem et nulla umbra defensam graviore aestu ardescere*). For the construction of the entrance-hall of the Domus Aurea, Via Sacra, one of the few streets in Rome with the title of *via* and so very important for traffic, was completely blocked; the street could not be used for Nero's triumphal procession in 68.[927]

So the inhabitants of Neropolis were not satisfied and the garden city of Nero did not exist for very long. During the reign of Vespasian, the area was again filled up with apartment buildings:

> *Deformis urbs veteribus incendiis ac ruinis erat; vacuas areas occupare et aedificare, si possessores cessarent, cuicumque permisit.*

> As the city was unsightly from former fires and fallen buildings, he allowed anyone to take possession of vacant sites and build upon them, in case the owners failed to do so.
>
> (Suet. *Ves.* 8.5)

Whoever wanted to build could do so. This meant: more tall *insulae*, more narrow lanes, more fire hazards. However, the lack of houses, still a problem in the time of Vespasian, was solved,[928] and the narrow lanes produced shadow. The situation regarding living and infrastructure before 64 was, for the most part, restored and the approach roads to Neropolis, for instance Via Flaminia or Via Appia, were probably not a lot different from those of Rome before and after Neropolis. Public space was further reduced by the construction of the Forum Pacis and the Colosseum, built in the same place as a former palace lake. The Domus Aurea, or what remained of it, was finally burnt down in 104.[929]

In the end, the government did not take the opportunity to create a clear street pattern with sufficient capacity to sustain traffic. One possibility for the reduction of traffic congestion would have been to remove Porta Trigemina. This gate, part of the Servian Wall, was a terrible obstacle at one of the busiest points of Rome, namely where the corn was brought from the granaries into the city. Nevertheless Porta Trigemina, possibly restored by Augustus notwithstanding, existed until the fourth century AD.[930]

A new chance: Colonia Commodiana

Shortly before the assassination of Commodus in 192, Rome was again burnt down by fire and a large part of the city was devastated. The fire started in or near the Templum Pacis on the Forum Pacis and struck other neighbouring areas. The fire was so severe that it could not be extinguished by anyone. Our main sources are Herodianus and Dio Cassius:

> *καταφλέξαν δὲ τὸ πῦρ τόν τε νεὼν καὶ πάντα τὸν περίβολον, ἐπενεμήθη καὶ τὰ πλεῖστα τῆς πόλεως καὶ κάλλιστα ἔργα· ὅτε καὶ τῆς Ἑστίας τοῦ νεὼ καταφλεχθέντος ὑπὸ τοῦ πυρὸς γυμνωθὲν ὤφθη τὸ τῆς Παλλάδος ἄγαλμα … κατεφλέχθη δὲ καὶ ἄλλα πλεῖστα τῆς πόλεως μέρη καὶ κάλλιστα, ἱκανῶν τε ἡμερῶν πάντα ἐπιὸν τὸ πῦρ ἐπεβόσκετο, οὐδὲ πρότερον ἐπαύσατο, πρὶν ἢ κατενεχθέντες ὄμβροι ἐπέσχον αὐτοῦ τὴν ὁρμήν.*

> After gutting the temple and its entire precincts, the fire spread to a major portion of the city, burning down some magnificent

buildings. The temple of Vesta, too, went up in flames on this occasion, exposing the statue of Pallas to view. The image, reputedly brought from Troy by the Romans, is venerated and kept concealed. Our generation were the first ones to see it since it had been brought to Italy from Troy. For the Vestal Virgins snatched up the image and carried it along the Sacred Way from the temple to the palace of the emperor. Very many other fine parts of the city were destroyed as the fire raged for several days, consuming everything in its way, and was only finally stopped when the rain began to fall and checked the blaze.

(Hdn. 1.14.4-5)

πῦρ τε νύκτωρ ἀρθὲν ἐξ οἰκίας τινὸς καὶ ἐς τὸ Εἰρηναῖον ἐμπεσὸν τὰς ἀποθήκας τῶν τε Αἰγυπτίων καὶ τῶν Ἀραβίων φορτίων ἐπενείματο, ἔς τε τὸ παλάτιον μετεωρισθὲν ἐσῆλθε καὶ πολλὰ πάνυ αὐτοῦ κατέκαυσεν, ὥστε καὶ τὰ γράμματα τὰ τῇ ἀρχῇ προσήκοντα ὀλίγου δεῖν πάντα φθαρῆναι. ἀφ' οὗ δὴ καὶ τὰ μάλιστα δῆλον ἐγένετο ὅτι οὐκ ἐν τῇ πόλει τὸ δεινὸν στήσεται, ἀλλὰ καὶ ἐπὶ πᾶσαν τὴν οἰκουμένην αὐτῆς ἀφίξεται. οὐδὲ γὰρ κατασβεσθῆναι ἀνθρωπίνῃ χειρὶ ἠδυνήθη, καίτοι παμπόλλων μὲν ἰδιωτῶν παμπόλλων δὲ στρατιωτῶν ὑδροφορούντων, καὶ αὐτοῦ τοῦ Κομμόδου ἐπελθόντος ἐκ τοῦ προαστείου καὶ ἐπισπέρχοντος. ἀλλ' ἐπειδὴ πάντα ὅσα κατέσχε διέφθειρεν, ἐξαναλωθὲν ἐπαύσατο.

And a fire that began at night in some dwelling leaped to the temple of Pax and spread to the storehouses of Egyptian and Arabian wares, whence the flames, borne aloft, entered the palace and consumed very extensive portions of it, so that nearly all the State records were destroyed. This, in particular, made it clear that the evil would not be confined to the City, but would extend over the entire civilized world under its sway. For the conflagration could not be extinguished by human power, though vast numbers both of civilians and soldiers carried water, and Commodus himself came in from the suburb and encouraged them. Only when it had destroyed everything on which it had laid hold did it spend its force and die out.

(D.C. 73.24.1-3)

Commodus also wanted to rename Rome, this time as Colonia Commodiana.

Fuit praeterea ea dementia, ut urbem Romanam coloniam Commodianam vocari voluerit.

> He had, besides, an insane desire that the city of Rome should be renamed Colonia Commodiana.
>
> (*Historia Augusta, Commodus* 8.6)

> Κομμοδιανὴν γοῦν τήν τε 'Ρώμην αὐτὴν.
> He actually ordered that Rome itself should be called Commodiana.
>
> (D.C. 73.15.2)

Commodus is seen as an infamous emperor, not explicitly as an arsonist like Nero; but nevertheless he was, according to the *Historia Augusta*.[931] Pertinax and his successors had the task of rebuilding Rome. Evidence of the bad condition of the roads in the Empire (probably also in Rome, after the fire) is given by the fact that Pertinax – although he had to save drastically after the reign of Commodus – raised funds to repair them.[932]

Nevertheless, the question remains whether Pertinax and his successors improved the infrastructure of Rome; probably the former situation was again restored. Caracalla built a new street (Via Nova)[933] from his Baths of Caracalla to the semi-circular side of the Circus Maximus. The street ran parallel to Via Appia, but it is improbable that this street was intended to relieve Via Appia of its traffic. It would far sooner have been an avenue to impress the people. The street must have had a considerable width, like a *platea*.[934]

As far as we know it is nowhere confirmed, in any source, that Rome had a metamorphosis during and after Commodus' reign like there was under Nero, by acquiring the name of Colonia Commodiana.[935] Coins with the name *Col*[onia] *L*[ucia] *A*[urelia] *Com*[modiana] already appear in 190.[936] The name Colonia Commodiana would not have meant that Rome had to be transformed into a Hippodamic *colonia* like Xanten or Trier, but should rather be translated as 'Commodus' own *colonia*'; Rome was indeed not a *colonia*, but the mother city of all *coloniae* in the Empire. According to the *Historia Augusta*, Commodus had the habit of renaming institutions and also cities after himself.[937]

A missed chance: the pomerium *as circular road*

An infrastructural measure often used nowadays is a ring road: a road encircling the centre of a city or the whole city, aiming to lighten the traffic in the centre. Nowadays, even villages sometimes have the disposition of a circular road.

The practice of diverting traffic flow around the centre of a city was not completely unknown. Rhodes and Alexandria had wide streets or avenues, intended to divert harbour traffic (mainly goods transport) around the centre.[938] Was the idea of diversion also adopted in Roman cities?

Every city founded by the Romans had a *pomerium*: the line which indicated the boundary of the city and its political entity. The Romans adopted the habit of the Etruscans in founding a city which went as follows: two white cattle, led by two *augures*, ploughed a furrow through the field where the boundary of the city was to come; a bull on the outside, a cow on the inner side of the furrow. The plough was lifted where the gates were to be.[939] From an archaeological point of view, the *pomerium* was also the city boundary consisting of the city wall with fallow land on either side which it was forbidden to plough or build upon:

> *Hoc spatium, quod neque habitari neque arari fas erat, non magis quod post murum esset quam quod murus post id, pomerium Romani appellarunt.*
>
> This space, which the gods forbade men to inhabit or to till, was called 'pomerium' by the Romans, quite as much because the wall stood behind it as because it stood behind the wall.
>
> (Liv. 1.44.5)

The *pomerium* was therefore holy ground, consecrated by *augures*, which had to remain unused.[940] In the aforementioned citation, use by traffic was not forbidden. It was possible and probably also permitted to enter the *pomerium*; in times of war, the defence structure had to be accessible to the defenders at every point. Because the *pomerium* surrounded the entire city, on both sides of the wall, paving it would have been an excellent way of conducting traffic around the city or its centre.

Did the *pomerium* function as a whole or in part circular road in Pompeii? In the original situation, when Pompeii was an Oscian settlement and no larger than the later districts VII and VIII, the through-road Naples–Nuceria ran along the city's boundary.[941] Here we have a situation where a road ran *alongside* the city and not *straight through* it. The road was parallel to the city wall, and when there was a *pomerium* in the Oscian situation the road had to pass the city either very closely or even over the *pomerium*.[942] In Figure 4.13 it can be seen that there was a circular road inside the defence wall; so in Oscian times the infrastructure was more extensive than later. After the Greek and Samnitic city extensions, the through-road became part of the street pattern of Pompeii and fulfilled the function of main street; the section running north–south acquired the function of *cardo*, (nowadays) Via Stabiana.

In this later situation, the use of the *pomerium* itself for traffic purposes does not seem very probable,[943] for the following reasons:

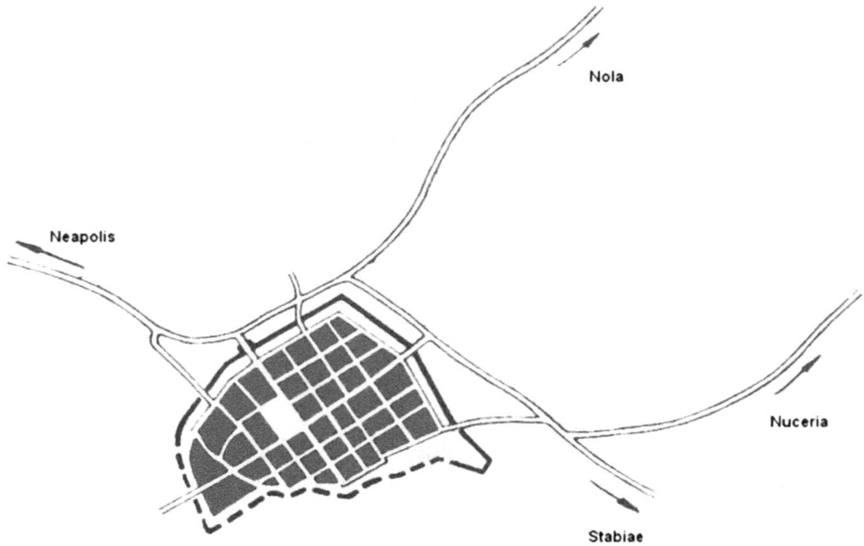

Figure 4.13 Pompeii, original Oscian settlement with circular road. Picture: C.R. van Tilburg.

1. The *pomerium* was a consecrated area; it was forbidden to use it for profane purposes, for instance as a street.[944] There was no pavement.
2. No wheel ruts are found in the *pomerium*.
3. The *pomerium* is interrupted at several points by gates and towers, giving entrance to the defensive wall. It was therefore possible to surround the city only along the wall, but for animals and wheeled traffic this was, of course, unfit for use. One has to assume, therefore, that the *pomerium* had no importance for traffic; at the most, the *pomerium* made it possible to reach the next *insula* or district on foot.

So, in Pompeii, the *pomerium* did not have any function for (wheeled) traffic and there was no circular road for traffic. What was the situation in cities founded later, especially the *coloniae* with their Hippodamic street pattern?

In Xanten a street has been found following the *pomerium*, inside the wall, surrounding the city.[945] Although large parts of Xanten have not yet been excavated, we have to assume that this street, like the other streets of Xanten, was wide enough to sustain traffic flow in both directions. The street follows the defensive walls and also the *pomerium*, which has a considerable width (Fig. 4.14).

So in Xanten there was a circular road, following the city side of the walls. At one point, this circular road is interrupted: at the point where the immunity of the Hafentempel reaches Hafentor (perhaps the circular road

Figure 4.14 Xanten, street along the *pomerium* at the city side of the wall. Photo: C.R. van Tilburg.

was also interrupted at the amphitheatre). Traffic driving around the centre had to pass the temple on the south-west side. The immunity of the temple was of more importance than the extension of the circular road. (Such a situation, where the immunity of a temple had more importance than the building of a profane object, is also visible on the Acropolis of Athens: the presence of the temple of Athena Nike made it impossible to construct the gate complex of the Propylaea as a symmetrical building.) Elsewhere, streets or roads followed part of the *pomerium*, as in Nîmes.[946]

It could even be possible that, during the planning, account had to be taken of traffic which ran north-west to south-east, parallel to the *cardo*. In the excavated part of the city, some streets do not run exactly parallel with *cardo* and *decumanus*: for example the street south-west of the amphitheatre and certain streets in the northern corner of the city, between *insulae* 22, 23, 29 and 30. These streets date from the period before Xanten acquired its status of *colonia*. It would have been possible to build over them, like the former north-west to south-east connection, where later *insulae* 15–21 were built, to the west of the new *cardo*. The former road was maintained east of the new *cardo*, creating obtuse corners.[947] The parallel route could have been as shown in Figure 4.15. The route along the *pomerium* has one awkward point to pass: the intersection between *insulae* 30, 31 and 36. Traffic, for which passing here was difficult, was able to use the longer route along *insulae* 22,

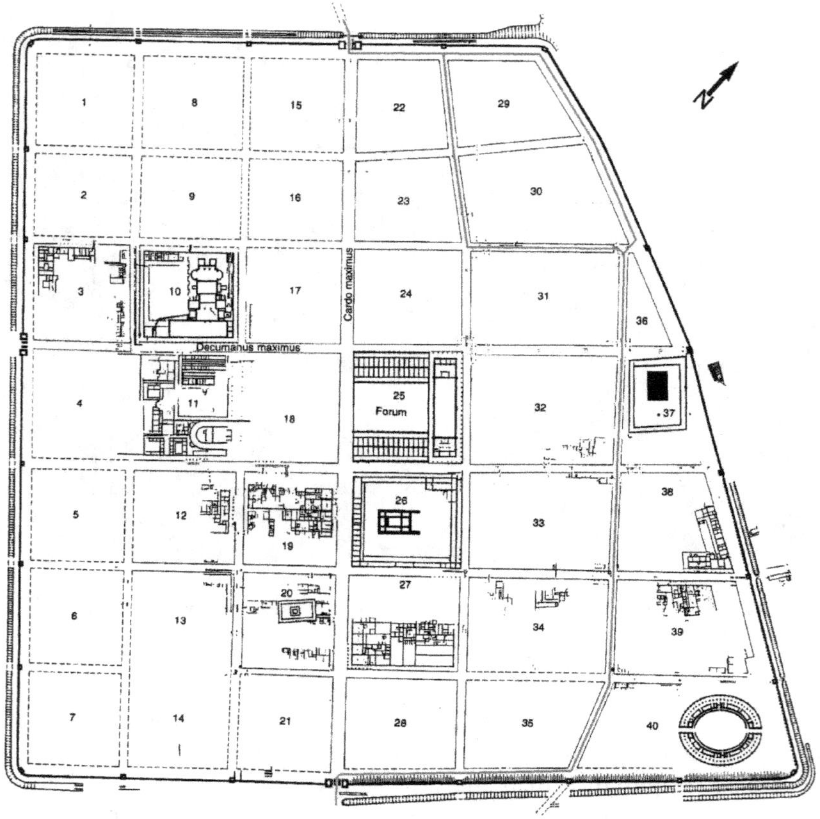

Figure 4.15 Xanten, parallel route. Heimberg and Rieche 7. H. Stelter, APX/C.R. van Tilburg.

23, 29 and 30. Most traffic, preferring obtuse angles, would have followed the '*pomerium* route'.

The street to the left of the amphitheatre runs at a slant, ending up at the wall tower. These towers are positioned at places where the side streets end at the *pomerium* and are always the same distance in relation to each other. To avoid a possible slant in the direction of the streets, the surveyors could have shortened the plots of *insulae* 29–35; these *insulae* now have a more rectangular shape than *insulae* 1–28. If a merely Hippodamic street pattern had been planned, the Hafentempel should have been projected a little more to the left, enabling the *pomerium* to continue round. In that situation, there would have been no obtuse corners at all.

One reason why the choice was made for this traffic-friendly situation would have been that here there was very intensive north-west to south-east traffic flow. The *cardo* was part of the *limes*, giving protection against the

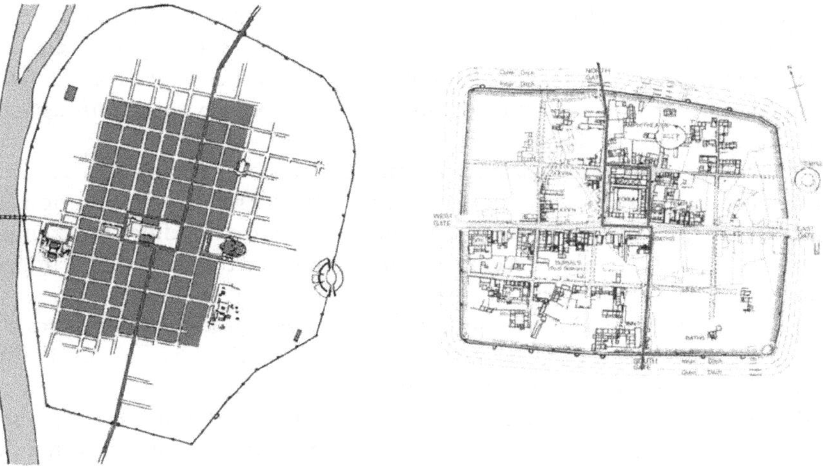

Figure 4.16 Diverted *cardines* in Trier and Caerwent. Left picture: C.R. van Tilburg. Right picture: Wacher 1997, 380. London: Routledge.

Germanic tribes, and there must have been many troop movements to and from the *castra* Castra Vetera (south-east of the city) and *castella* such as Asciburgium and Burginatium. As well as that, there was enough space to create a city with wide streets.

In other cities patterns are sometimes also visible, with regard to certain routes, where there is an intervention in the straightness of both *cardo* and *decumanus*. In Pompeii as well as in Xanten, *cardo* and *decumanus* are straight streets running from gate to gate (over the full length, however, the *decumanus* of Pompeii shows a slight bend). It could also occur, however, that this was not the case and a bend is visible. This is to be found in Trier, Silchester (original situation), St Albans and Caerwent (Fig. 4.16). In Tongeren and Silchester the through-roads (in both cities the *decumani*) were diverted but in Trier, St Albans and Caerwent they were maintained. The building of a new *cardo* in a city like Caerwent would have been very radical, but the reason why the bend was maintained in Trier is not absolutely clear; the gates could easily have been erected exactly opposite each other, creating a straight *cardo*, able to cope with busy traffic. An explanation could be that here fixed routes were also in use; the *forum* would then possibly have functioned as a type of roundabout.

Why these measures were taken in Tongeren and Silchester could have had to do with the improvement of the traffic circulation: to separate the flow of local and interurban traffic around the forum, where the traffic was very busy. At the time when the diversions were created, Tongeren and Silchester were open cities, without a wall, so the creation of diversions was relatively simple.[948] Another point is that obtuse corners are scarce or absent.[949] A pure

Hippodamic design also caused the absence of obtuse parallel roultes as in Xanten; in the case of the *forum* as roundabout there was no issue here of a busy central crossing in the city. So the streets could therefore have gradually attained another function; they could have been 'promoted' to throughfare street, or 'demoted' to local street.

The traffic situation in *castra* and *castella* was comparable with Xanten. They all had a Hippodamic street pattern and circular roads were common.[950] Tacitus uses the word *ambitus* here, usually relating to a building (for example a temple).[951] Cities which developed from military settlements, like Timgad, usually have an excellent infrastructure.

In Rome, the situation was quite different. Even in the earliest times the city had a *pomerium*, which was extended at a later stage by Servius Tullius or, according to Tacitus, perhaps even by Titus Tatius.[952] Information on both the course of the *pomerium* and its width in the Republican era is not known, but it presumably followed the Servian Wall. The next statesman of whom it is said that he extended the *pomerium* is Sulla,[953] followed later by Caesar.[954] Then the Servian Wall had already been dismantled; the Temple of Venus Genetrix at the Forum Julium is even built over the trace of the former wall,[955] and during the time of Horace one could walk over the former wall.[956] The wall as fortification no longer existed, but the *pomerium* was still a sanctified place. Some emperors, for instance Claudius, Vespasian and Hadrian,[957] extended the *pomerium*; the largest expansion was done by Aurelian, who erected the new wall.[958] On the other hand, Palmer states that the course of the Aurelian Wall was already fixed as tax boundary by Vespasian (in 73–74) or even Augustus.[959] Should this have been the case, this boundary would not have followed the *pomerium*, fixed by Aurelian's predecessors.

The *pomerium* of Rome was not indicated by strips of fallow land and there was certainly no circular road. Instead of that, the *pomerium* was marked by stones known as *cippi*, with an inscription mentioning the emperor who extended the *pomerium*.[960] *Cippi* are also found in Capua[961] and Varro mentions *cippi* in Aricia.[962]

Was it possible to create a circular road around Rome? Considering the building of the Aurelian Wall, surrounding the entire city, this question can be answered in the affirmative.[963] A circular road could have solved much of the traffic congestion; traffic with the other side of the city as destination could then have used the circular road. Nevertheless, such a circular road was never built. We also have to conclude here that the government did not see the importance of planning and building such a circular road. From a financial point of view also, circular roads outside the tax area would not have been attractive: there was tax on all incoming traffic.[964]

Circular roads were still an exception in early modern times. The facsimile edition of Dutch cities of Frederick de Wit (*c.* 1700) shows that by far the majority of cities in the northern and southern Netherlands (Netherlands, Belgium and some cities in northern France) did not have the disposition of a circular road. In Britain in *c.* 1800 they were also absent.[965]

Political lack of interest

Finally, the question arises as to why a people, technically able to construct straight roads, bridges, city walls and aqueducts, as well as pure Hippodamic cities and agricultural areas, never made the effort to improve deficient infrastructure and instead always confined themselves to introducing laws and traffic-restricting measures. Nowhere was a road or street widened to improve the traffic flow, a building pulled down which was in the way, an intersection adapted, or a circular road constructed, although it was technically possible. What was the reason for this passive attitude to traffic congestion?

In our mobile society, where traffic plays a very large role in daily life, it is perhaps difficult to imagine that there were periods in history when this was not the case. Not because the volume of traffic was smaller than nowadays, but because traffic had a lower priority. In Roman cities the military-defensive aspect, the religious dimension and the problem of water management, both the supply of drinking water and the removal of sewage and excess rainwater, played a larger role than the management of traffic flow. Of course it was important that supplies to cities was maintained, but as long as the cities were still accessible and functioning the local government did not consider it necessary to take measures to improve the traffic situation, making it easier to approach and to drive through cities. So the absence of circular roads and wide streets was a result of a mental attitude.

Apart from that, the governing officials – local, provincial and imperial – generally belonged to the political upper class, far removed from the common people.[966] This upper class spent its time – apart from their governmental tasks – on agriculture (when they were off duty) and had a downright disdain for trade, industry and finance. Under the terms of the *Lex Flaminia*, senators were not even allowed to concern themselves with trade.[967] But it was exactly the artisans, tradesmen and merchants who suffered the most from traffic congestion: they were directly responsible for the supply of goods to the cities.

The lack of interest in trade, industry and finance of the agriculture-loving upper class is strikingly described by Cicero in his following citation.

> *Iam de artificiis et quaestibus, qui liberales habendi, qui sordidi sint, haec fere accepimus. Primum improbantur ii quaestus, qui in odia hominum incurrunt, ut portitorum, ut faeneratorum. Illiberales autem et sordidi quaestus mercennariorum omnium, quorum operae, non quorum artes emuntur; est enim in illis ipsa merces auctoramentum servitutis. Sordidi etiam putandi, qui mercantur a mercatoribus,*

quod statim vendant; nihil enim proficiant, nisi admodum mentiantur; nec vero est quicquam turpius vanitate. Opificesque omnes in sordida arte versantur; nec enim quicquam ingenuum habere potest officina. Minimeque artes eae probandae, quae ministrae sunt voluptatum: 'Cetarii, lanii, coqui, fartores, piscatores', ut ait Terentius; adde huc, si placet, unguentarios, saltatores totumque ludum talarium.

[151] *Quibus autem artibus aut prudentia maior inest aut non mediocris utilitas quaeritur, ut medicina, ut architectura, ut doctrina rerum honestarum, eae sunt iis, quorum ordini conveniunt, honestae. Mercatura autem, si tenuis est, sordida putanda est; sin magna et copiosa, multa undique apportans multisque sine vanitate impertiens, non est admodum vituperanda, atque etiam, si satiata quaestu vel contenta potius, ut saepe ex alto in portum, ex ipso portu se in agros possessionesque contulit, videtur iure optimo posse laudari. Omnium autem rerum, ex quibus aliquid acquiritur, nihil est agri cultura melius, nihil uberius, nihil dulcius, nihil homine libero dignius.*

Now in regard to trades and other means of livelihood, which ones are to be considered becoming to a gentleman and which ones are vulgar, we have been taught, in general, as follows. First, those means of livelihood are rejected as undesirable which incur people's ill-will, as those of tax-gatherers and usurers. Unbecoming to a gentleman, too, and vulgar are the means of livelihood of all hired workmen whom we pay for mere manual labour, not for artistic skill; for in their case the very wage they receive is a pledge of their slavery. Vulgar we must consider those also who buy from wholesale merchants to retail immediately; for they would get no profits without a great deal of downright lying; and verily, there is no action that is meaner than misrepresentation. And all mechanics are engaged in vulgar trades; for no workshop can have anything liberal about it. Least respectable of all are those trades which cater for sensual pleasures: 'Fishmongers, butchers, cooks, and poulterers, and fishermen,' Terence says. Add to these, if you please, the perfumers, dancers, and the whole *corps de ballet.* But the professions in which either a higher degree of intelligence is required or from which no small benefit to society is derived – medicine and architecture, for example, and teaching – these are proper for those whose social position they become. Trade, if it is on a small scale, is to be considered vulgar; but if wholesale and on a large scale, importing large quantities from all parts of the world and distributing to many without misrepresentation, it is not to be greatly disparaged. Nay, it even seems to deserve the highest respect, if those who are engaged in it, satiated, or rather, I should say, satisfied with the fortunes they

> have made, make their way from the port to a country estate, as they have often made it from the sea into port. But of all the occupations by which gain is secured, none is better than agriculture, none more profitable, none more delightful, none more becoming to a freeman.
>
> (Cic. *Off.* 1.150–1)

Cities had to function by the transport of goods in and out. It was good enough if there was an infrastructure for trade and industry, even if the majority of that infrastructure was set up as one-way streets. Should a shopkeeper, a baker or greengrocer complain about the situation, the local government probably did not listen to him. The removal of excess rainwater and sewage, the city defence and the presence of religious immunities were of higher priority, because these affected the ruling class directly. Also more important was city decoration: the position of statues and imposing buildings along the thorough-fares and streets. This aspect, only important for the upper class, must have stimulated the local government to conduct as much traffic as possible past these monuments[968] and kept it from building circular roads.

A law such as the *Lex Julia Municipalis* only reinforced the negative image of people such as merchants, artisans and shopkeepers. A man, working for money, was better working at night, as far as possible from the houses of the upper class. A noble man, a so-called *homo liber* as Cicero says, had another occupation at night, which was to sleep.

Summary and conclusion

Did the Romans really try to improve the traffic situation by enlarging the infrastructure? Generally speaking, it can be stated that in a city the existing infrastructure did not change significantly. Also, when cities were destroyed by fire, the brutality of war or natural disasters, changing the street pattern was an exception. Even the infamous fires in Rome, during the reigns of Nero and Commodus, did not ultimately result in improvements to the former traffic chaos; Nero even aggravated the situation further by closing Via Sacra.

Technically, cities must have been able to build circular roads, but because of the religious immunity of the *pomerium* this never became reality. One also had to keep account of the fact that the local government did not want to create a circular road for passing traffic: they wanted to create an impression on passing travellers.

Traffic was merely incidental. City embellishment (Neropolis!), water management, religion and city defence were more important. Infrastructure was only important when it did not interfere with other things and there was enough space, as in Xanten.

Conclusion of chapter 4

In this last chapter I have discussed how Roman politicians, both at a local as well as a state level, attempted to solve traffic congestion, or at any rate to keep it under control.

This took place first of all by legislation against all forms of driving traffic: wheeled traffic and ridden animals. Transferring certain types of goods transport to the nocturnal hours by means of the *Lex Julia Municipalis* solved the problem only partially, because, for example, construction traffic and traffic for the disposal of waste products (detritus, rubble and excrement) was still permitted by day. Neither was the abolishment of horse-ridden traffic sustainable, as demonstrated by the repeated laws of later emperors. Furthermore, in the third century driving with wagons in the densely populated and chaotic Rome was again permitted to senators.

The erection of physical barricades in daily traffic was more successful. As a consequence, wheeled traffic had to take stipulated routes. In Pompeii, the best-preserved Roman city, excavations show that traffic had to take fixed routes: routes where the streets were wide enough and where obtuse corners in streets with one-way traffic facilitated the passing of wheeled traffic. By means of the introduction of barricades, the city council could alter routes. In Pompeii worn wheel ruts have been found in streets later barricaded by stones.

Regarding the presence of deeper wheel ruts on obtuse corners one can conclude that wagon drivers preferred these turnings and, perhaps, this was even a traffic regulation. This is also found in Hippodamic Xanten, where a worn cornerstone of a harbour gate gives evidence of extensive traffic flow which passed this obtuse bend.

Modern solutions, such as enlargements of infrastructure to cope with traffic, did not occur in the Roman Empire. When a city was planned, the street pattern remained the same; striking changes did not take place. Even in the case of devastation, when the city government had the chance to learn from former congestion and to start again with a *tabula rasa* by building a new, larger and better infrastructure, they did not take this apportunity. In some cases, the infrastructure was even more restricted, for instance the interruption of Via Sacra by Nero's Domus Aurea. The city government considered traffic of least importance, a matter of less importance than the removal of waste water, the maintenance of defence structures and religious immunities (the *pomerium* and the immunity of temples).

This can be compared with our traffic policy in old historic city centres nowadays. Today streets are also blocked by traffic cones or barricades; there are certain hours when trucks are permitted to load or unload; and cars are forced to use fixed one-way routes.

SUMMARY

The Romans were not the first to construct paved metalled carriageways – that had already taken place long before the Romans, *c.* 3000 BC in Mesopotamia – but they were the first who constructed an intricate road-system, reaching the most distant outlying areas. For road-users to gain as much time as possible – in the first place the army and the many kinds of messengers – the roads were built, where possible, in a straight line. To obtain these straight lines, natural barriers – such as rivers, mountains, valleys, marshes, deserts and sometimes even a strait – had to be crossed by bridges and even tunnels. Generally, the roads were kept in an excellent state of repair and facilities were available everywhere along the roads, in the form of milestones and inns (including facilities for changing horses) for long-distance travellers. One can state that the level of road construction and management only exceeded that of the Romans for the first time in the twentieth century, as a result of motorised vehicles and the application of asphalt and reinforced concrete.

In Antiquity, not everyone was able to travel. However, individuals could travel, either on foot, by ass or horse or in a wagon, but it was expensive, time-consuming and sometimes also dangerous – which was the case not only in Antiquity, but for the whole preindustrial age. Only the arrival of first the steam train, then the affordable car and (in recent decades) cheaper plane tickets have made it possible for large groups of people to travel and to live further away from office and workplace, resulting in traffic congestion in the morning and evening rush-hours, public transport and mass tourism. In Antiquity, these phenomena were of course unknown. The traffic of which we are the best informed is the state transport company (the *cursus publicus*), the army and emperors on inspection or in the field. This traffic flow was, however, not very large, and moreover they travelled mainly over long distances on interurban highways, causing relatively little traffic congestion. The largest type of traffic flow, without any doubt, was anonymous goods transport, supplying the cities daily. It was this traffic which caused the greatest congestion and against which measures were particularly introduced.

Because of the well-surfaced roads and the relatively small numbers of road-users on interurban roads, the chance of delay would have been very small. The only delay apart from road repair was created by the toll points which one had to pass. There were relatively many toll points, particularly in mountainous areas. Traffic congestion was mainly limited to urban areas. The traveller approaching a city surrounded by a wall had first to pass the city gate. Older cities in particular usually had narrow single gates, through which only one horseman or wagon could pass at a time. Later, gates were erected with two, three or four passages, which could cope with considerably more traffic. Nevertheless, as a result of legislation, traffic was forced to accept restrictions: a traveller wishing to cross a city by day had to wait until the evening, unless he had permission to do so. Messengers in the service of the emperor, for example, would have had such a dispensation. Wagons would not have been convenient in the Roman Empire: they were expensive and subject to restrictions. If one wished to travel to another city, it was in many cases better and cheaper to rent a wagon with driver at the city gate. If it was very busy, someone could even go before the wagon to create space.

The traffic laws which have come down to us, of which the *Lex Julia Municipalis* is the most important, mainly had the task of dividing pedestrian, riding and wheeled traffic. There was already a division in pedestrian and riding traffic in the form of sidewalks, sometimes colonnaded, but goods traffic, both entering and leaving, had to move by night, as stated in the *Lex Julia Municipalis*. By day, there was still plenty of wheeled traffic in the city, caused by the exceptions, with permission under the *Lex Julia Municipalis*.

Control of traffic was further realised by introducing barricades in order to force traffic to take a certain route. This situation occurred especially in older cities with streets mainly too narrow for intersecting or passing traffic. These barricades were applied in the first instance to wheeled traffic, but horses and mules would also have suffered. Stepping stones, typical of the street scenery of Pompeii, also caused traffic hindrance and dictated the width of the axles of vehicles. In other, more northerly cities, wider streets were constructed which could sustain more traffic, probably with fewer barricades.

In this book about traffic circulation two cities have been more researched, each considered as an example of an 'old' and a 'new' city, respectively Pompeii and Xanten. Pompeii is a city which dates initially from the sixth century BC and was enlarged twice. Some main streets are wide enough to have allowed traffic to pass or to overtake each other, but most streets are too narrow and could only have functioned as one-way routes. This resulted in a system whereby traffic had to follow stipulated routes, based on the presence of barricades and obtuse corners. The gates were also generally narrow and did not permit more than one wagon at a time entering or leaving. The traffic circulation is comparable to that of an old Dutch city centre, where there are also fixed routes and times for traffic, for example Leiden, or in England Ipswich.

Xanten is a city founded later, *c.* 100 AD, and planned with a Hippodamic street pattern in a spacious manner. There was neither the need nor the desire to build narrow streets, and the presence of much military traffic created the need for wide streets. So even by Roman standards, the streets were wide. Nevertheless, archaeological research gives evidence that here also fixed routes were in use. Traffic congestion was, however, minimal and not comparable with the problems of Pompeii. Xanten had three large city gates, able to contend with much traffic. This city is comparable with a modern Dutch city with specially equipped streets, for example Lelystad, or in Britain Milton Keynes.

The third researched city, Rome, is another case entirely. One can correctly state that Rome is the most un-Roman city of the Roman Empire. The city had an extremely chaotic street plan, caused by high land prices, uncontrolled jerry-building and the presence of hills, which was far from sufficient in keeping traffic flow under control. Apart from some approach roads, Rome was a labyrinth of lanes and blind alleys, the majority too narrow for wagons. As far as we know, attempts to improve the traffic situation had no results, not even after devastating fires in 64 and 192.

The question remains whether traffic congestion and traffic restricting measures were perceived as a large problem, like today. Roman society was basically different in many ways. For us, traffic congestion is a problem, because we do not have enough time. Even in our free time we want to be on the beach, in the forest or by the lake as soon as possible, in our attempts to miss as little pleasure as possible. In Antiquity, time was not yet measured with clocks. Furthermore, there were far fewer travellers, therefore also far fewer people who were hindered by traffic congestion, and for many travellers traffic restricting measures were not applicable to them.

The only people who really suffered from traffic congestion and restrictive measures were individual travellers (including people who travelled as a profession) using wagons or riding animals, and goods transporters, who had to supply the cities. This last group particularly belonged to a social class with which the local government did not have much affinity. As long as traffic was able to circulate, however laboriously, there was no problem. Traffic was merely incidental, of lesser importance than defence, fine views, city ornamentation, water management and religious immunities.

NOTES

1. For example Statius' panegyric of Via Domitiana between Sinuessa (south-east of Rome) and Naples (*Silvae* 4.3).
2. For pre-industrial empires see Claessen 164–6.
3. Wegener Sleeswyk 4.
4. Forbes 1934, 73–80 and App. 1.
5. X. *An.* 1.5.7 *Καὶ δή ποτε στενοχωρίας καὶ πηλοῦ φανέντος ταῖς ἁμάξαις δυσπορεύτου.*
6. Forbes 1934, 80–3; Chapot 779. Herodotus (5.52–3) describes the course of the road, including fortresses, stages and distances.
7. Forbes 1934, 50–5.
8. Chapot 778.
9. Knauss 329. In this book, I usually give dimensions in feet and miles in a classical context; in other contexts, I usually give them in metres and kilometres. A foot is *c.* 30 cm; a (Roman) mile is *c.* 1.6 km.
10. Knauss *passim.*
11. Forbes 1934, 101–3; E. *El.* 775 *δίκροτον* [. . .] *ἁμαξιτόν*. Wheels with flanges following a fixed track were already being built in the fifth century BC at the Diolkos (ship portage) of Corinth: Bulle 7, 11, 28–9.
12. White 92.
13. For an extended enumeration for the Greek terms see Chapot 777–8.
14. Str. 5.3.8 *ὠλιγώρησαν ἐκεῖνοι, στρώσεως ὁδῶν* [. . .] *οἱ δ'* (Roman) *ὑπόνομοι συννόμῳ* [. . .] *ὁδοὺς ἁμάξαις χόρτου πορευτὰς ἐνίας ἀπολελοίπασι*; Forbes 1934, 96.
15. Paus. 8.54.5 (a well-paved road): 'Η δὲ ἐς Ἄργος ἐκ Τεγέας *ὀχήματι ἐπιτηδειοτάτη καὶ τὰ μαλιστά ἐστι λεωφόρος*, but this can be a Roman road, cf. Chapot 810–11; (a badly paved road) 10.5.5 'Η δὲ λεωφόρος *αὐτόθεν ἡ ἐς Δελφοὺς καὶ προσάντης γίνεται μᾶλλον καὶ ἀνδρὶ εὐζώνῳ χαλεπωτέρα*; for more citations of Pausanias concerning bad roads see Forbes 1934, 100. Liv. 37.7.13 *pontes in fluminibus factos, vias, ubi transitus difficiles erant, munitas vidit*; White 93 (who *falso* refers to Liv. 7.37).
16. Ar. *V.* 259 *οὑτοσί μοι βόρβορος φαίνεται πατοῦντι*. For more examples of Greek and Hellenistic cities with unpaved roads see Forbes 1934, 108.
17. Forbes 1934, 109; *IG* II^2 I 1–2, 1191, 21–22; Zimmermann 184.
18. Plu. *Moralia* (*Praecepta gerendae reipublicae*) 811 B (Epaminondas) *προήγαγε τὴν τελεαρχίαν, οὐδὲν οὖσαν πρότερον ἀλλ' ἢ περὶ τοὺς στενωποὺς ἐκβολῆς κοπρίων καὶ ῥευμάτων ἀποτροπῆς ἐπιμέλειάν τινα*. Kudlien (p. 42) is using the term 'Telmarch'.
19. Arr. *An.* 1.26.1 *ᾗ ὡδοπεποιήκεσαν αὐτῷ οἱ Θρᾷκες* [. . .] *μακρὰν οὖσαν τὴν πάροδον* (Alexander). Why especially Thracians were responsible for this work is not clear to me; in my opinion, it was an arbitrary group of soldiers in Alexander's army.
20. Forbes 1934, 107–8.

21. Zimmermann 182, with a reference to the 'Astynomeninschrift von Pergamon', *SEG* 13 521 (see Klaffenbach, 1954).
22. Str. 14.1.37 *αἱ ὁδοὶ λιθόστρωτοι.*
23. Forbes 1934, 111.
24. J. *BJ* 1.425 *τὴν δ' Ἀντιοχέων τῶν ἐν Συρίᾳ πλατεῖαν οὐ φευκτὴν οὖσαν ὑπὸ βορβόρου κατέστρωσέν τε, σταδίων εἴκοσι τὸ μῆκος οὖσαν, ξεστῇ μαρμάρῳ καὶ πρὸς τὰς τῶν ὑετῶν ἀποφυγὰς ἐκόσμησεν ἰσομήκει στοᾷ.*
25. Kudlien 42–4 with references.
26. Serv. *A.* 1.422 *Strata viarum. Primi enim Poeni vias lapidibus stravisse dicuntur*; Isid. *Orig.* 15.16.6 *Primi autem Poeni dicuntur lapidibus vias stravisse*; Forbes 1934, 115–16; Chapot 783.
27. D.S. 20.44.5 *διὰ τῶν στενωπῶν*; Forbes 1934, 116.
28. Forbes 1934, 116–17 and 121; Martha 254; Chapot 783 n. 12 states that the Carthaginians transported the knowledge of road building (with the use of mortar) from Egypt to Italy; this should confirm the theory that Roman road building has Carthaginian roots. Forbes 1950, 75 (= 1952, 56) mentions the Etruscans, Greeks and Carthaginians as teachers of the Romans.
29. Forbes 1934, 117 and 121.
30. Forbes 1934, 117–18.
31. Forbes 1934, 118–21.
32. Forbes 1934, 119; Chapot 783 especially nn. 8–11.
33. Quilici 1999, 21–6.
34. Forbes 1934, 119; Giebel 131–2.
35. Giebel 132.
36. Cozzo 40–1.
37. Liv. 5.40.10 *L. Albinius* [. . .] *virgines sacraque in plaustrum imposuit et Caere, quo iter sacerdotibus erat, pervexit.*
38. Liv. 2.11.7 *Gabina via.*
39. Liv. 2.39.4 *in Latinam viam* [. . .] *transgressus.*
40. Liv. 7.9.6 *Salaria via.*
41. About the year 312 see Liv. 9.29.5 *Et censura clara eo anno Ap. Claudi et C. Plauti fuit; memoriae tamen felicioris ad posteros nomen Appi, quod viam munivit*; Pekáry 40 with, in n.16, the proposed year 310. Quilici (1999, 27) quotes that *munivit* does not mean that a *via strata* was realised; in that case, the word *stravit* should be used.
42. Amongst others D.S. 20.36.2 *μετὰ δὲ ταῦτα τῆς ἀφ' ἑαυτοῦ κληθείσης Ἀππίας ὁδοῦ τὸ πλεῖον μέρος λίθοις στερεοῖς κατέστρωσεν ἀπὸ Ῥώμης μέχρι Καπύης*; Anon. *De viris illustribus* 34.6 (p. 78) *Appius Claudius Caecus* [. . .] *Viam* [. . .] *stravit, unde illa Appia*; Humm 735 especially n. 97.
43. Humm 694; Radke 1964–75, 1243–4.
44. Humm 694; Pekáry 37–46.
45. Liv. 10.23.12 *semitamque saxo quadrato a Capena porta ad Martis straverunt*; Humm 699.
46. Liv. 10.47.4 *via a Martis silice ad Bovillas perstrata est*; the Loeb text mentions the year 292; Humm 699. The year 342, mentioned by André and Baslez (p. 390), seems to me improbable.
47. Liv. 41.27.5 *censores vias sternendas silice in urbe, glarea extra urbem substruendas marginandasque primi omnium locaverunt*; Humm 696.
48. Humm 713.
49. Liv. 7.39.16 *viae quae nunc Appia est perveniunt*; Humm 713 n. 42.
50. Rome–Aricia (23 km) and Genzano–Terracina (61 km); Humm 705 and 719–24.
51. Humm 722–4; he states that the Proposition of Pythagoras was already known in Rome in the fourth century BC. The Babylonians already had knowledge of the Proposition of Pythagoras (Sarton 1 73) and it is possible that this Proposition had reached Italy

via others. For the use of the Proposition in geodesy see Clavel-Lévêque 161–84 *passim*, especially 172–3; here the course of a pre-Roman road in the neighbourhood of Béziers is discussed, where trigonometry is used.

52. Cic. *Tusc.* 4.2.4 *Mihi quidem etiam Appii Caeci carmen, quod valde Panaetius laudat epistola quadam, quae est ad Q. Tuberonem, Pythagoreum videtur.*
53. *CIL* X 6824–35; D.C. 68.15.3[1] Καὶ κατὰ τοὺς αὐτοὺς χρόνους τά τε ἕλη τὰ Πομπτῖνα ὡδεποίησε λίθῳ (the Loeb text mentions the year 110); Chapot 795 (in the year 107); Humm 703-4. This shows that the citation of Anon. *De viris illustribus* 34.6 *Appius Claudius Caecus* [. . .] *Viam usque Brundisium lapidibus stravit, unde illa Appia* is wrong.
54. About Capua as terminus see amongst others D.S. 20.36.2 *supra*; Fron. *Aq.* 5 (Appius) *qui et Viam Appiam a Porta Capena usque ad urbem Capuam muniendam curavit*; Humm 711–13.
55. For an overview of the roads including year of commencement, politician and course from the realisation of the Via Appia onwards to the middle of the second century BC see Hinrichs 164–5. For the road building in the period 133–109 BC see Hinrichs 168–70.
56. Pekáry 49; Hinrichs (p. 164) does not deny 225 BC as being the year of beginning.
57. Hinrichs 164.
58. Pekáry 49; Hinrichs 164.
59. According to Pekáry, the year of beginning of Via Cassia is completely unknown (p. 49). Hinrichs (p. 165) gives as possible years 154 and 125 BC.
60. Casson 164; according to Pekáry, this is not certain (p. 49). Hinrichs (p. 165) mentions 187 BC.
61. Chapot 783 and 794.
62. Pekáry 48; Hinrichs (165) mentions 188 BC.
63. Liv. 39.2.7–10 *M. Aemilius alter consul* [. . .] *viamque a Placentia, ut Flaminiae committeret Ariminum perduxit.* In the same year the other consul, C. Flaminius, realised a road from Bologna to Arrezzo; Liv. 39.2.6 *viam a Bononia perduxit Arretium.*
64. Plb. 34.12.2[a] Ἐκ δέ τῆς Ἀπολλωνίας εἰς Μακεδονίαν ἡ Ἐγνατία ἐστὶν ὁδὸς. Casson dates the road 148 BC (p. 164).
65. Cic. *Font.* 18 *ad rem publicam pertineret viam Domitiam muniri*; Chapot 784.
66. App. *BC.* 1.23 ὁ δὲ Γράκχος καὶ ὁδοὺς ἔτεμνεν ἀνὰ τὴν Ἰταλίαν μακράς; Chapot 783.
67. Plu. *Caes.* 5.5 ἐπεὶ δὲ τοῦτο μὲν ὁδοῦ τῆς Ἀππίας ἀποδειχθεὶς ἐπιμελητὴς πάμπολλα χρήματα προσανάλωσε τῶν ἑαυτοῦ; Chapot 784.
68. Hinrichs 170.
69. Forbes 1952, 57; Rogge (p. 102) mentions 372 main roads with a total length of about 85,000 km. Von Stephan (p. 70) gives the numbers as 75,000–80,000 km.
70. Bender 1983, 141.
71. The width of roads will be discussed in detail in chapter 1, pp. 26–31.
72. *Codex Theodosianus* 15.3.4; van Binnebeke 93–4.
73. Claudius Claudianus, *Carmina minora* 40.7–8 *quin tempore nullo | cessant Flaminiae pulverulenta viae*; Zimmermann 194.
74. Rutilius Namatianus 1.39–41 *Postquam Tuscus ager postquamque Aurelius agger | perpessus Geticas ense vel igne manus | non silvas domibus, non flumina ponte coercet. Agger* has here the significance of 'road', referring to the bed of the road. This word also gives the image of a road of which the surface has disappeared.
75. Procop. *Goth.* 1.14.6 Ὁ δὲ διὰ τῆς Λατίνης ὁδοῦ ἀπῆγε τὸ στράτευμα.
76. Esch 5. See also *infra.*
77. Rogge 103.
78. Davies 108. The illustration shows that wheel ruts in the Middle Ages, caused by bad maintenance, were far deeper than in Antiquity.

79. Hindle 6.
80. Liv. 41.27.5; see *supra.*
81. Amongst others Fest. p. 241M (p. 276 Lindsay) *Publicius clivus appellatur, quem duo fratres L. M. Publici Malleoli aediles curules pecuaris condemnatis ex pecunia, quam ceperant, munierunt, ut in Aventinum vehiculi<s> †hel venire† possit*; Var. *L.* 5.158 *Clivos Public< i>us ab aedilibus plebei Publici<i>s qui eum publice aedificarunt*; Stambaugh 188.
82. Platner and Ashby 232.
83. Liv. 5.55.2–5 *promisce urbs aedificari coepta. Tegula publice praebita est; saxi materiaeque caedendae, unde quisque vellet ius factum praedibus acceptis eo anno aedificia perfecturos. Festinatio curam exemit vicos dirigendi, dum omisso sui alienique discrimine in vacuo aedificant. Ea est causa ut veteres cloacae, primo per publicum ductae, nunc privata passim subeant tecta, formaque urbis sit occupatae magis quam divisae similis.* For more information of the city planning of Rome see chapter 4.
84. For more information on the Servian Wall see chapter 3, pp. 91–4.
85. Liv. 41.27.7 *clivum Capitolinum silice sternendum curaverunt.*
86. Liv. 41.27.12 *Pisauri viam silice sternendam.*
87. Mommsen 1877, 486–7.
88. Mommsen 1877, 487; Gesemann 55–6 (Pompeii).
89. Gesemann 55–6.
90. Platner and Ashby 232–3.
91. Plin. *Nat.* 19.23 *Caesar dictator totum forum Romanum intexit viamque sacram ab domo sua et clivum usque in Capitolium* (all references to Pliny the Elder are those of the Loeb edition of Rackham; if not, the other edition is mentioned); Platner and Ashby 233–4.
92. Stambaugh 189; according to Mommsen, in 709 AUC (44–45 BC) not yet all streets were paved (1877, 489).
93. Var. *L.* 7.15 *leges iubent in directo pedum VIII esse <viam>, in anfracto XVI, id est in flexu*; Stambaugh 188. Given in feet 8 equals about 2.40 m. For the widths of roads see chapter 1, pp. 26–31.
94. Var. *L.* 5.22 *Via quidem iter, quod ea vehendo teritur*; *R.* 1.2.14 *rustici etiam nunc quoque viam veham appellant propter vecturas.*
95. For more kinds of *viae* see Chapot 782.
96. Ulp. *dig.* 43.8.2.22 *publicas vias dicimus, quas Graeci βασιλικάς, nostri praetorias, alii consulares vias appellant.* A *via consularis* or a *via praetoria* is, in fact, also a *via publica.*
97. Ulp. *dig.* 43.7.3.1–3 *Sed inter eas* [. . .] *intermoriuntur.* Here *viae militares* are seen as a category of *viae vicinales.* For the full text see n. 304.
98. *AE* 1922, 210; van Binnebeke 11–12.
99. See n. 79 *supra.*
100. Paul. *Fest.* p. 371M (p. 508 Lindsay) *Viae* [. . .] *privatae, quibus neminem uti* [. . .] *praeter eorum quorum sunt*; Ulp. *dig.* 43.8.2.22–23 *Privatae sunt* [. . .] *ducens.* Van Binnebeke 12 n. 33.
101. Paul. *Fest.* p. 371M (p. 508 Lindsay) *Viae sunt et publicae, per* [. . .] *e omnibus licet*; Ulp. *dig.* 43.7.1 *Cuilibet* [. . .] *interdicitur*; 43.8.2.21–22 *Viam publicam* [. . .] *vias appellant*; Chapot 782. For an overview of all conditions for a *via publica* see Pekáry 6, even mentioned by Rathmann (p. 411). For other criteria for a *via publica* see Rathmann 412–13 (Ulpianus).
102. Ulp. *dig.* 43.8.2.22 *Vicinales sunt viae, quae in vicis sunt vel quae in vicos ducunt*; Bender 1989, 109.
103. Paul. *Fest.* p. 17M (p. 16 Lindsay) *Actus* [. . .] *iter inter vicinos quattuor pedum latum*; Chapot 782.
104. Chapot 782.

105. Liv. 27.16.2 *dilapsique per nota urbis itinera*; van Binnebeke 12. Her reference to Liv. 9.2.9–10, however, refers to *viae*.
106. Fest. p. 6M (p. 5 Lindsay) *ambitus proprie dicitur circuitus aedificiorum patens in latitudenem pedes duos et semissem, in longitudinem idem quod aedificium*; cf. 17M (p. 15 Lindsay). *Thesaurus Linguae Latinae* s.v. 'ambitus' 1858.
107. Var. *L.* 5.22 *ab eoque Duodecim Tabularum interpretes 'ambitus parietis' circuitum esse describunt*; Isid. *Orig.* 15.16.13 *ambitus inter vicinorum aedifica locus, duorum pedum et semipedis ad circumeundi facultatem relictus, et ab ambulando dictus.*
108. Var. *L.* 5.35 *qua id anguste, semita, ut semiter dictum*; van Binnebeke 17 with commentary; Chapot 782. Because an *iter* has a width of 2 feet, a *semita* only has a width of 1 foot, about 30 cm.
109. Cic. *Sest.* 12 *calles et pastorum stabula praeoccupare coepisset*; Liv. 22.14.8 *deviasque calles exercitum ducimus conditi nubibus silvisque*; Serv. *A.* 4.405 *'callis' est semita tenuior, callo pecorum praedurata*; van Binnebeke 19; Laurence 1999, 145-6 (transhumance). For an extended overview of *calles* in Italy see Waldherr's article.
110. Var. *L.* 7.62 *Domum ire c<o>epi tramite <in> dextra via. Trames a transverso dictus*; van Binnebeke 19.
111. Ter. *Eu.* 635 *ubi ad ipsum veni devorticulum, constiti*; Fron. *Aq.* 5 *Concipitur Appia* [. . .] *deverticulo*; van Binnebeke 19; for a paved *diverticulum* see Laurence 1999, 56.
112. Some *clivi* are already mentioned: Clivus Publicius and Clivus Capitolinus; Cic. *Att.* 2.1.7 *equitatus* [. . .] *in clivo Capitolino*; van Binnebeke 19; Chapot 782.
113. Juv. 3.236–237 *raedarum transitus arto | vicorum in flexu*; Tac. *Ann.* 15.38 *flexis atque enormibus vicis, qualis vetus Romam fuit.* Van Binnebeke 19; Stambaugh 184.
114. Fest. p. 18M (p. 16 Lindsay) *Angiportus iter compendiarium in oppido, eo quod sit angustus portus, id est aditus in portum*; Var. *L.* 5.145 *Angiportum, si<ve quod> id angustum, <sive> ab agendo et portu.* Van Binnebeke 19; Harsh 1937 *passim.*
115. Var. *L.* 5.145 *Fundulae a fundo, quod exitum non habe<n>t ac pervium non est*; van Binnebeke 19.
116. Van Binnebeke 19.
117. Pl. *Am.* 1011 *nam omnis plateas perreptavi*; Isid. *Orig.* 15.2.23 *Plateae perpetuae ac latiores civitatum viae sunt*; van Binnebeke 18 with commentary; Harsh 1937 *passim.*
118. Isid. *Orig.* 15.2.24 *Quintana pars plateae quinta est, qua carpentum progredi potest.*
119. Homo 363.
120. Homo 372–6.
121. Ulp. *dig.* 8.3.pr. *Via est ius eundi et agendi* [. . .] *nam et iter et actum in se via continet*; A. Kolb 2000, 206.
122. Paul. *dig.* 8.3.7.pr. *qui actum habet* (beside the right of *iter*) *et plostrum ducere et iumenta agere potest.*
123. *Oxford Latin Dictionary* s.v. 'actus'.
124. Ulp. *dig.* 8.3.pr. *Iter est ius eundi ambulandi homini, non etiam iumentum agendi.*
125. Roman jurists always consider the real width of roads in practice, in cases concerning *via, actus* and *iter*; van Binnebeke 15.
126. Ulp. *dig.* 8.5.4.pr.-1 *Loci corpus non est dominii ipsius, cui servitus debetur, sed ius eundi habet. Qui iter sine actu vel actum sine itinere habet, actione de servitute utetur.*
127. *AE* 1981, 210 = *AE* 1922, 127 refers to a private person who realised a road, paid for by himself, suitable for wagons (*via plostralis*).
128. Paul. *dig.* 9.2.31 *quod si nullum iter erit, dolum dumtaxat praestare debet, ne immittat in eum, quem viderit transeuntem: nam culpa ab eo exigenda non est, cum divinare non potuerit, an per eum locum aliquis transiturus sit*; Chapot 782.
129. This system is also used in Dutch polders like Schermer, Zuidplas and Flevoland (although here there is talk of rectangular plots) and American cities.
130. In the (primary) literature, mostly the spelling *kardo* is used.

131. About the origin of the word *decumanus* see Dilke 88.
132. Dilke 87.
133. Fest. p. 116M (p. 103 Lindsay) *Limites in agris nunc termini, nunc viae transversae.*
134. *Actuarius* (van Binnebeke 18) is close to *actus*, a linear measure of 120 feet (Fest. p. 17M [p. 16 Lindsay] *in geometrica minorem partem iugeri, id est centum viginti pedum*) or 35.48 m. It is the distance which a team of oxen can track a plough (*agere*) until the point of return; Dilke 1971, 81.
135. Dilke 93 with illustrations *infra*.
136. Hyg. *de limitibus* p. 71 *ceteri autem limites, qui subruncivi appellantur, patere debent ped. VIII*; van Binnebeke 18.
137. Barth's text.
138. *Lex XII* tab. VII (p. 44) *viam muniunto.*
139. Van Tilburg 2002, 193.
140. Forbes 1934, 132–3; Rogge 107; Davies 55. The words *statumen, rudus, nucleus* and *pavimentum* are derived from Vitruvius (7.1.3 *Tunc* [. . .] *tesseris*), concerning mosaic floors.
141. Forbes 1934, 138.
142. Rogge 109.
143. Forbes 1934, 140.
144. Rogge 108.
145. Mertens 1997, 15–16.
146. There are countless examples. I will mention two: Cassino: Valenti 42–3; Rome: beside Via Appia also Via Nomentana: Quilici Gigli 1997 *passim*. In Britain, where the soil is less rocky and cities and settlements had to sustain less traffic, 80 per cent of the urban streets consisted of *viae glarea stratae*; Davies 63.
147. For a stratigraphy of eleven layers see Chapot 785.
148. Forbes 1934 *passim*; exceptions in Britain amongst others 148 and 149.
149. Meyer mentions a road not constructed immediately along a river in danger of inundation; it is the road along the Rhine in Switzerland (p. 137).
150. One can find a map of Roman roads and river valleys between the Po region and Augsburg in Cartellieri (Übersichtplan, without page number).
151. Meyer 129; Forbes 1934, 146.
152. Str. 4.6.6 *προσέϑηκε γὰρ ὁ Σεβαστὸς Καῖσαρ τῇ καταλύσει τῶν ληστῶν τὴν κατασκευὴν τῶν ὁδῶν, ὅσην οἷόν τ' ἦν*; Meyer 121 and 128.
153. Cartellieri 49–52; for more information on Via Claudia Augusta see Czysz 101–16 and Grabherr 117–30.
154. Cartellieri 94–5; Forbes 1934, 146.
155. Cartellieri 96–7.
156. Forbes 1934, 146.
157. Str. 4.6.7 about the accessibility of the passes in the West Alps region: *καὶ ἡ μὲν διὰ τοῦ Ποινίνου λεγομένου φέρεται, ζεύγεσιν οὐ κατὰ τὰ ἄκρα τῶν Ἄλπεων, ἡ δὲ διὰ Κευτρώνων δυσμικωτέρα*; 4.6.11 *διττὴ δ' ἐστίν, ἡ μὲν ἁμαξεύεσϑαι δυναμένη διὰ μήκους πλείονος, ἡ διὰ Κευτρώνων, ἡ δὲ ὀρϑία καὶ στενή, σύντομος δὲ ἡ διὰ τοῦ Ποινίνου*. After that he enumerates the four passes from Italy to Gaul. On passes not passable for wagons see also Forbes 1934, 146; Meyer 122 and 139. He mentions a wide road of 7–8 feet (2.10–2.40 m), suitable for wagons, on p. 130.
158. Forbes 1934, 140–1, 145–6; Chapot 786. For ruts in the Balkan region: Bulle 30–4 and Tafel 3. In the Alps natural and artificial ruts occur: Bulle 37–8 and *passim*. For ruts in Switzerland: Herzig 2002, 11–16 and G. Schneider 67–74, and in France (Alsace): G. Schneider, 75–7.
159. Walser 65–8; for more information and illustrations see Barocelli 18–40.

160. Johnston 85; Davies 81.
161. Johnston 79 and 139; Davies 81. Davies has made an extended study of gradient roads in Britain; for gradient percentages see Davies 79–86. For a reconstruction picture of the stairway in front of the south gate in Lincoln see p. 134.
162. Str. 4.6.6 οὐ γὰρ δυνατὸν [. . .] ἐν τοῖς ἡλίοις.
163. Claudius Claudianus, *De bello Gothico* 321–328 (Stilicho) *ocius* [. . .] *curat.*
164. Forbes 1934, 40–4; Hindley 11.
165. See for this theory chapter 1, pp. 33–4.
166. Tac. *Ann.* 1.63 *Caecina qui suum* [. . .] *antevenisset.*
167. Forbes 1934, 45.
168. Tac. *Ann.* 1.61 *pontesque et aggeres umido paludum et fallacibus campis inponeret.* This is the evidence that a *pons* and an *agger* are different constructions.
169. Radke 1903–78, 1444.
170. Catul. 17.1–4 *O Colonia, quae cupis ponte ludere longo,* | *et salire paratum habes, sed vereris inepta* | *crura ponticuli axulis stantis in redivivis,* | *ne supinus eat cavaque in palude recumbat.* In my opinion, this is not a *pons longus* as described above, but merely a long wooden bridge.
171. Colenbrander 2005, 69; see p. 68 for other swamp road constructions found in the Netherlands.
172. Van Buren 2428.
173. Forbes 1934, 151.
174. Rogge 109. For the proportions of paved and unpaved sections of this road in Limburg see Bloemers 250–2 and Table 1.1.
175. Forbes 1934, 140–1 with illustrations.
176. Plu. *CG* 7.2 ἄλλους δὲ λίθους ἔλαττον ἀπέχοντας ἀλλήλων ἑκατέρωθεν τῆς ὁδοῦ διέθηκεν, ὡς εἴη ῥᾳδίως τοῖς ἵππους ἔχουσιν ἐπιβαίνειν ἀπ' αὐτῶν ἀναβολέως μὴ δεομένοις.
177. Adams 51. My explanation for this is that they were simple stones which could be removed easily.
178. Radke 1903–78, 1438.
179. Plu. *CG* 7.2 διαμετρήσας κατὰ μίλιον ὁδὸν πᾶσαν [. . .] κίονας λιθίνους σημεῖα τοῦ κατέστησεν; Forbes 1934, 151.
180. H.-C. Schneider 105; *CIL* I^2 21 = X 6838.
181. Bender 1989, 119; H.-C. Schneider 103.
182. Plin. *Nat.* 3.66 *a miliario in capite Romani fori statuto ad singulas portas*; Adams and Laurence 48.
183. H.-C. Schneider 107.
184. Singor 55.
185. Mommsen 1877, 490 with references to *CIL* III 479 and 6093.
186. Cartellieri 157.
187. H.-C. Schneider 106.
188. The length of a Roman mile (*milia passuum*) is 1481.5 m; H.-C. Schneider 102.
189. H.-C. Schneider 102 n. 1. The length of a *leuga* is about 1.5 Roman miles or about 2200 m; Ammianus Marcellinus 16.12.8 *quarta leuga* [. . .] *et decima id est unum et viginti milia passuum*: Schulzki 99.
190. For example *AE* 1984, 889–91.
191. Pl. *Hipparch.* 228 D (Hipparchus) ἔστησεν αὐτοῖς Ἑρμᾶς κατὰ τὰς ὁδοὺς ἐν μέσῳ τοῦ ἄστεος καὶ τῶν δήμων ἑκάστων; Eitrem 701; for more information on Greek milestones see Pekáry 61–7 and André and Baslez 379–81.
192. Rogge 110–11.
193. Nowadays, a 'viaduct' is a bridge, crossing another road. In secondary literature, when

the word 'viaduct', 'viaduc' or 'viadotto' is used, the significance is commonly roads on earthen walls or dam roads, for example Rottier 32 and Forbes 1934, 51.

194. Van Buren 2428 and 2430.
195. Arch constructions of aqueducts were called *opus arcuatum*; for example Fron. *Aq.* 14 *ex eo opere arcuato* [. . .] *passuum trium milium septuaginta sex* (about the Aqua Claudia); van Buren 2429.
196. Van Buren 2432.
197. Van Buren 2432; Radke 1903–78, 1438.
198. Van Buren 2432 and 2433.
199. *ILS* 5898 *Imp. Caes. Nerva | Traiano Aug. Ger | Dacico pont.max. | trib. pot. cos. V p.p. | Aquiflavienses | pontem lapideum | de suo f.c.* We have to assume that the explicit quotation of *lapideum* gives the evidence of a former wooden bridge. For more inscriptions see van Buren 2433.
200. For an extended overview of all roads in the Roman Empire which still exist see Gazzola.
201. Cuijk: Colenbrander 2005, 69 (over the Meuse); Cologne: *CCAA Romeins Keulen* 16 and 36; van Tilburg 2004, 395 (over the Rhine). They date both from the fourth century AD.
202. For an enumeration of recently found bridges in Britain see Davies 90–3.
203. For example to the island Djerba off the coast of Africa (Tunisia).
204. Esch 100; Laurence 1999, 198. For more retaining wall constructions in the region of Rome see Esch 76, 77 and 114.
205. Mertens 1997, 14–18, also giving numbers of other bridges.
206. Quilici Gigli 1997, 47–53.
207. In Bonomi Ponzi's article, some culverts are described in Via Flaminia; the culvert of Pieve Fanonica on p. 159. For British culverts see Davies 98–104.
208. Quilici 1997 *passim*.
209. For all constructions in the Dora valley west of Aosta see Barocelli 17–50 with many illustrations and Mollo Mezzena 57–72.
210. Hdt. 3.60 ὄρεός τε ὑψηλοῦ ἐς πεντήκοντα καὶ ἑκατὸν ὀργυιάς, τούτου ὄρυγμα κάτωθεν ἀρξάμενον, ἀμφίστομον. τὸ μὲν μῆκος τοῦ ὀρύγματος ἑπτὰ στάδιοι εἰσί, τὸ δὲ ὕψος καὶ εὖρος ὀκτὼ ἑκάτερον πόδες. [. . .] ἀρχιτέκτων δὲ τοῦ ὀρύγματος τούτου ἐγένετο Μεγαρεὺς Εὐπαλῖνος Ναυστρόφου. A στάδιον had a length of 185 m and a Greek foot or πούς 31 cm; a little longer than a Roman foot. A Roman mile was a little bit longer than 8 *stadia*. Thus Eupalinos' tunnel had a length of about 1.3 km.
211. About the pre-Roman history of tunnel construction see Grewe 33–69.
212. Grewe 135–9.
213. Grewe 124–6; Bender 1989, 117; *CIL* X 1614 *L · COCCEIVS · L · | C · POSTVMI · L · | AVCTVS ARCITECT* (*sic*).
214. For the course of the roads in and around Naples see Basso *et al.* 117, 119 and 123.
215. Str. 5.4.5 τοῦ Κοκκηίου τοῦ ποιήσαντος τὴν διώρυγα ἐκείνην τε καὶ ἐπὶ Νέαν πόλιν ἐκ Δικαιαρχείας ἐπὶ ταῖς Βαίαις.
216. Grewe 125–6; he uses the spelling 'cripta'.
217. Grewe 126. According to Bender (1989, 117) it dates from Claudius' time.
218. Str. 5.4.7 ἔστι δὲ καὶ ἐνθάδε διῶρυξ κρυπτή, τοῦ μεταξὺ ὄρους τῆς τε Δικαιαρχείας καὶ τῆς Νεαπόλεως ὑπεργασθέντος ὁμοίως, ὥσπερ ἐπὶ τὴν Κύμην, ὁδοῦ τε ἀνοιχθείσης ἐναντίοις ζεύγεσι πορευτῆς, ἐπὶ πολλοὺς σταδίους.
219. Grewe 127.
220. *CIL* XI^2 6106; Chapot (p. 787) dates it in the year 77 and gives the following dimensions: length 37 m, width 4–6 m and height 4–5 m.
221. Grewe 129–34.

222. Forbes 1934, 143.
223. Chevallier 1997, 153.
224. Giuliani 34 (Percorso B).
225. For gradient percentages west of Tivoli see Giuliani 34–5.
226. Giuliani 37; Boëthius and Ward-Perkins 140.
227. Boëthius and Ward-Perkins 140.
228. Giuliani 34–5 (Percorso A).
229. Santillo Frizell 35.
230. Var. *L.* 7.15 *leges iubent in directo pedum VIII esse <viam>, in anfracto XVI, id est in flexu*; van Binnebeke quotes still more citations on p. 27 of jurists maintaining these dimensions.
231. Gaius *dig.* 8.3.8 *Viae latitudo ex lege duodecim tabularum in porrectum octo pedes habet, in anfractum, id est ubi flexum est, sedecim*; van Binnebeke 27–32.
232. Paul. *Fest.* p. 371M (p. 508 Lindsay) *Et ita privatae VIII pedes in latitudine iure et lege* [...] *lex iubet XVI <in anfracto fle>xuque pedes esse vias ut qui* [...] *vias muniunto: †onisam† dilapidassunt, qua volet, iumento ageto*; van Binnebeke 27–9.
233. Paul. *Fest.* p. 371M (p. 508 Lindsay) *publicae quantum ratio utilitatis permittit*; van Binnebeke 27.
234. Paul. *dig.* 8.3.23.pr. *Via constitui vel latior octo pedibus vel angustior potest*; van Binnebeke 27 and 31.
235. On the widths of *viae privatae* amongst others *CIL* II 3443 *ITER · PRIVAT | AB · PARIETE|N · ITER · LATE|PEDES · III* (3 feet); *CIL* V 3473 *ITER|T · VIBI · T · L · |ERONIS|P · IISLA* (2.5 feet). A *via privata* can, however, be wider: *CIL* XI^2 5042 *A · X · VIR · S IN · P · XX ... |P · XXIV · LA XIS · INI · P · XXXV|LA · XV · IN · P · XXXIV · LA XVS|IN · P · IIII · LA XVS · IN · P · XI · LAT ... |XXIV · EX · INDE· VIA · PVB.* For a *via privata* like a *via publica*, see Cic. *Q. fr.* 3.1.4 *viam perspexi, quae mihi ita placuit, ut opus publicum videretur esse*; Laurence 1999, 54 and 103–4.
236. Paul. *Fest.* p. 17M (p. 16 Lindsay) *Actus* [...] *iter inter vicinos quattuor pedum latum*; Chapot 782. See the section 'Development of the Roman road-system' *supra* s.v. 'actus' and 'iter'.
237. Javol. *dig.* 8.3.13.2 *Latitudo actus itinerisque ea est, quae demonstrata est: quod si nihil dictum est, hoc ab arbitro statuendum est. in via aliud iuris est: nam si dicta latitudo non est, legitima debetur*; van Binnebeke 31.
238. For 30, 15, 12 and 8 feet: Hyg. *De limitibus* p. 71 *Maximus decimanus et cardo plus patere debent sive ped. XXX, sive ped. XV, sive ped. XII, sive quot volet cuius auctoritate fit. Ceteri autem limites, qui subruncivi appellantur, patere debent ped. VIII*; for 40, 20, 12 and 8 feet (incl. the *actuarii*): Hyg. *Constitutio <limitum>* p. 157 *decimano maximo pedes XL, kardini maximo pedes XX, actuariis* [*autem*] *limitibus omnibus decimanis* [*et*] *kardinibus pedes XII, subruncivis pedes VIII*; van Binnebeke 23.
239. For the exact relationship of these numbers to each other see van Binnebeke 23–4.
240. *Liber Coloniarum* 1, p. 212 *Qui conduxerit, decimanum latum ped. XL, kardinem latum p. XX facito, et a decimano et kardine m. quantum quenque facito ped. XII, ceteros limites subruncivos latos ꝯ ii facito. m = maximo*; ꝯ ii means, probably, 8: van Binnebeke 26 n. 90.
241. Plin. *Nat.* 17.169 *vineas limitari decumano XVIII pedum latitudinis ad contrarios vehiculorum transitus, aliisque traversis limitibus denum pedum distingui per media iugera, aut, si maior modus sit, totidem pedum cardine quot decumano limitari, semper vero quintanis semitari, hoc est ut quinto quoque palo singulae iugo paginae includantur.* The *quintana* followed always the fourth vine; that is the origin of the name. A *quintana* also existed in a military settlement: Plb. 6.30.6 ἦν [...] παρήκειν.
242. Hyg. *Constitutio <limitum>* p. 157 *Limitibus latitudines secundum legem et constitutionem divi Augusti debemus*; hereafter the width dimensions of note 238 *supra* follow; van Binnebeke 25–6; Pekáry 22.
243. H.-C. Schneider 19; Bender 1989, 109; Pekáry 35.

244. Pekáry 35.
245. Pekáry 22–3; Grénier 1934, 366–7; Quilici 1999, 24, 30 and 31 (Italy: many paved road sections of 4.1 m or about 13 feet).
246. Casson 171; according to Friedländer (pp. 277–8) 4.30 m (14 feet).
247. *CIL* III 10773 *PVB · VIA · LATA · P.L*; Pekáry 22.
248. Casson 171; Bulle (Taf. 16 and 17) shows some road profiles with standard rut widths of 0.93 m. Taf. 17 (i. Klais) shows a profile of 1.50 m (5 feet). This profile took part of the Brenner Pass route; Klais is situated in the neighbourhood of Partenkirchen. For a road width of 5 feet see also Meyer 122 and Bulle 31 (Dalmatia). Pekáry notes 2.50–5 m (8–16 feet) as width measurement for roads in Switzerland.
249. Hyg. *De limitibus* p. 144 *Quibusdam coloniis decumanum maximum ita constituerunt, ut viam consularem transeuntem per coloniam contineret; sicut in Campania coloniae Axurnati*; Pekáry 23–7.
250. Forbes 1950, 74–5 (= 1952, 56).
251. Rogge 109; van Es 107.
252. Rogge 109. On the Via Appia, shepherds with herds were prohibited: Forbes 1934, 128.
253. Davies 71. For widths of paved road sections in Britain, Davies 71–5.
254. The information in this table is derived from Bloemers 250–2; Stuart and de Grooth 17–18, 22 and 26 and Rogge 109 give slightly different dimensions. For more road widths in north-west Europe see Bender 1989, 116 n. 42.
255. Grénier 1934, 366.
256. Schultze 298 and Taf. XIII.
257. See chapter 3, pp. 98–9.
258. Zimmermann 182; Klaffenbach 4–8 col. 1, 23–9 *ἔστωσαν* [...] *χρῶνται* (Astynomeninschrift von Pergamon, *SEG* 13 521).
259. Forbes 1934, 110–11; Rottier 53. For more detailed street widths see Hoepfner and Schwandner ill. 186 (Priene).
260. Str. 17.1.8 *ἅπασα μὲν ὁδοῖς κατατέτμηται ἱππηλάτοις καὶ ἁρματηλάτοις, δυσὶ δὲ πλατυτάταις, ἐπὶ πλέον ἢ πλέθρον ἀναπεπταμέναις, αἵ δὴ δίχα καὶ πρὸς ὀρθάς τέμνουσιν ἀλλήλας*. A *πλέθρον* = 100 feet, 31 m. For the real dimension of Canopus Street see Hoepfner and Schwandner (p. 239), mentioning 30 m (100 feet); *contra* Empereur 56, mentioning a width of 14 m. For the widths of Canopus Street and other streets see Hoepfner and Schwandner 238–40. Colonnades: Ach. Tat. 5.1.2 *στάθμη μὲν κιόνων ὄρθιος ἑκατέρωθεν ἐκ τῶν Ἡλίου πυλῶν εἰς τὰς Σελήνης πύλας* [...] *ἐν μέσῳ δὴ τῶν κιόνων τῆς πόλεως τὸ πεδίον*.
261. In later times, they were narrowed and built up; Hoepfner and Schwandner 20, 21 and 308.
262. Zimmermann 185; Hoepfner and Schwandner 306. Roads to harbour basins in Rhodes and Alexandria: Hoepfner and Schwandner 67 and ill. 41 and 225.
263. Mau 221; H.-C. Schneider 82. Etienne (1985, 299) gives as exact dimensions in metres: Via dell'Abbondanza 8.47–8.53 m with sidewalks of 3.96–4.30 m; Via di Nola 7.26–8.36 m with sidewalks of 3.54 m; Via Stabiana 7.15–7.47 m with sidewalks of 3.98 m. He also quotes that the other streets have widths of between 3 and 5 m. For even more detailed width dimensions of the main streets see Nissen 543–8 (in metres and feet).
264. Such a width corresponds with the 8 feet norm of the *viae publicae*. In a city like Pompeii, where the presence of higher sidewalks (also occupying space) rendered giving-way impossible, only one-way traffic could take place.
265. The traffic circulation in Pompeii, I will discuss in chapter 4, pp. 137–43.
266. According to Platner and Ashby, there were two Viae Novae: a Via Nova at Forum Romanum, on the *Forma Urbis* (Lugli and Gismondi; Lanciani 29) known as Nova Via, and a Via Nova running from the Baths of Caracalla to the Circus Maximus (Lugli and Gismondi; Lanciani 42; Platner and Ashby 565.

267. Stambaugh 188. According to Platner and Ashby (p. 568) there were two Viae Tectae; both had a colonnade. The first was situated in the Campus Martius: Sen. *Apoc.* 13 *per campum Martium; et inter Tiberim et viam tectam descendit ad inferos*; Mart. 8.75.2 *a Tecta Flaminiaque.* The second was outside Porta Capena: Ov. *Fast.* 6.192 *Tectae porta Capena viae.*
268. Homo 363.
269. F. Kolb 418.
270. The *Forma Urbis* (Lugli and Gismondi) mentions Via Triumphalis; the Arch of Constantine is built over this street.
271. Pekáry 36.
272. H.-C. Schneider 82; he quotes 2.90 m as minimum width.
273. See chapter 4.
274. Hull 64–75; Forbes 1934, 156; for more street width information see Davies 77-8 (also in feet).
275. Carmelez 11.
276. Rieche 21.
277. Mommsen 1871–88 2^1, 454; Pekáry 46.
278. Liv. *Perioche* 20 <*C. Flaminius censor viam Flaminiam*> *muniit*; Chapot 783.
279. Hinrichs 167.
280. Radke 1964–75, 1243-4; Pekáry 46–53; *contra* Chapot 787; H.-C. Schneider 47–50. See also notes 43 and 44 *supra* about Via Appia.
281. Pekáry 49–50 and 53; Radke 1903-78, 1434.
282. Ulp. *dig.* 43.8.2.21 *Viam publicam eam dicimus, cuius etiam solum publicum est*; Pekáry 53.
283. Pekáry 51; see note 47 *supra.*
284. For a complete enumeration of the censors' activities concerning road building and maintenance see Radke 1903–78, 1433–6.
285. Chapot 783.
286. *ILLRP* 454a *CCLX* | *T.Annius T.f.* | pr[*aetor*]; Radke 1903–78, 1433.
287. Radke 1903–78, 1433.
288. Chapot 783–4.
289. *ILLRP* 1277 [*C.?A*]*urelius* | [*L.?f.*] *Cottas* | [*c*]*onsol* | *LVII*; *contra* Pekáry 46 n. 43 about the date.
290. *CIL* I^2 17 *APPIOS CONSOL.*
291. *CIL* I^2 617 *M · AEMILIVS · M · F · M · N*|*LEPIDVS · COS*; 618 *M AEMILIVS · MI · M · N*|*LEPID · COS.*
292. *CIL* I^2 619 *M · AEM*|*LEPID.*
293. *CIL* I^2 624 *S · POSTVMIVS · S · F · S · N*|*ALBINVS · COS.*
294. *CIL* I^2 654 *L · CORNELIV*|*S · L · f · CINA · COS.*; Radke 1903–78, 1432 no. 9 (Via Latina).
295. *CIL* I^2 661 *L · CAECILIQ · F*|*METEL · COS.*
296. According to Chapot (p. 798) this road ran from Sinuessa to Cumae; there was a connection with the tunnel road to Naples discussed in chapter 1, p. 23–5. *CIL* X pp. 58 and 702.
297. Chapot 798.
298. Cic. *N.D.* 3.11 *videmus deos* [. . .] *in Salaria Vatinius.*
299. Cic. *Font.* 18 *cum ad rem publicam pertineret viam Domitiam muniri, legatis suis, primariis viris, C. Annio Bellieno et C. Fonteio, negotium dedit.*
300. Cic. *Man.* 55 *Appia iam via carebamus.*
301. Radke 1903–78, 1426–7. Via Flaminia got its name from the end of the third century AD onwards; Staccioli 17.
302. Eck 1979, 25; Finley 126.

303. In any case, the army played a role in the realisation of Via Flaminia: Liv. 39.2.6 *ne in otio militem haberet, viam a Bononia perduxit Arretium.*
304. Ulp. *dig.* 43.7.3.1–3 *Sed inter eas et ceteras vias militares hoc interest, quod viae militares exitum ad mare aut in urbes aut in flumina publica aut ad aliam viam militarem habent, harum autem vicinalium viarum dissimilis condicio est: nam pars earum in militares vias exitum habent, pars sine ullo exitu intermoriuntur.*
305. Pekáry 12, 13 and 36; H.-C. Schneider 21–3.
306. Chapot 782.
307. *Liber Coloniarum* 1, p. 241 *per certa loca viae militares finem faciunt* (Dalmatia); *CIL* III 6123 (Nero) *TABERNAS · ET · PRAETORIA|PER · VIAS · MILITARES|FIERI IVSSIT.* (Thracia); Cic. *Prov.* 4 *via illa nostra, quae per Macedoniam est usque ad Hellespontum militaris* and *Pis.* 40 *obsessio militaris viae* (Via Egnatia).
308. Suet. *Aug.* 49.3 *iuvenes primo modicis intervallis per militaris vias, dehinc vehicula disposuit.*
309. Pan. Lat. 5.7.2 *militaris vias ita[c] confragosas et alternis montibus arduas atque praecipites*; André and Baslez 397–8.
310. Pekáry 12–13.
311. Von Petrikovits I, 54; Giebel 140 (Agrippa).
312. The different phases to unlock new areas is described by Johnston (pp. 50–1); such a pattern is also visible in development countries nowadays. In development countries, however, there is already a population before the construction of such roads; in the Roman Empire, on the other hand, the infrastructure was realised first and the colonies were founded later.
313. *CIL* I^2 638 (Lucania); Hinrichs 170–2. According to Hascher and Sonnabend (p. 23) Via Popilia in the Po valley.
314. Liv. 27.37.9 *aedilium curulium* [. . .] *intraque decimum lapidem ab urbe*; Pekáry 55–6. *Contra* the *Lex Julia Municipalis* 68–70 *Quae loca publica* [. . .] *procuratio est* and van Binnebeke 124, who state that these powers reached no further than one mile outside the city. The fact that the powers extended outside the city can be proved by the fact that outside the city infrastructural constructions were also built, like the widening roads approaching gates and aqueducts.
315. *CIL* III 14202^4 *Floru* [*s q*]; Pekáry 68–9. So the addition is unclear.
316. *CIL* $I^1 593 = I^2 744$; van Binnebeke 47.
317. Cic. *Fam.* 8.6.5 (Curio) *legemque viariam* [. . .] *iactavit*; van Binnebeke 47–8; Pekáry 69–70.
318. Cic. *Att.* 9.9.2 *Arpinum* [. . .] *via deterrima*; 16.13a.2 *Appiam* [. . .] *via mala.*
319. Cic. *Att.* 1.1.2 *Thermus* [. . .] *curator est viae Flaminiae*; Chapot 787. The fact that Via Flaminia was a busy road with a bad surface was mentioned by Martial (9.57.1–5) *Nil est tritius Hedyli lacernis* [. . .] *nec quae Flaminiam secant salebrae*; Zimmermann 194.
320. *CIL* I^2 751 *L · FABRICIVS · C · F · CVR · VIAR|FACIVNDVM · COERAVIT*; Platner and Ashby 400.
321. D.C. 47.17.4 τὰς ὁδοὺς οἰκείοις οἱ βουλευταὶ δαπανήμασιν ἐπεσκεύαζον.
322. *CIL* X 6895, 6897, 6899–6901 (Via Latina); Pekáry 71–2. About Valerius Messal(l)a: Tib. 1.7.57–60 *Nec taceat monumenta viae* [. . .] *hic glarea dura* | *sternitur, hic apta iungitur arte silex.* Eck (1979, 28) calls him Messalla Corvinus.
323. D.C. 53.22.1 τὰς ὁδοὺς τὰς ἔξω τοῦ τείχους δυσπορεύτους ὑπ' ἀμελείας ὁρῶν.
324. Aug. *Anc.* 20 *Con(s)ul septimum viam Flaminiam a(b urbe) Ari(minum feci et pontes)* | *omnes praeter Mulvium et Minucium*; maybe the *curator viarum* C. Minucius Thermus was the builder of this bridge. *CIL* XI^1 365 *SENATVS · POPVLVSQ*[*ue romanus*] | *imp. caesari divi f. augusto imp. sept* |*COS · SEPT · DESIGNAT · OCTAVOM · V*[*ia flamin*]*IA* |*CELEBERRIMEIS · ITALIAE · VIEIS · CONSILIO* [. . .] (on the triumphal arch of Rimini).
325. Suet. *Aug.* 30.1 *Quo autem facilius undique urbs adiretur, desumpta sibi Flaminia via*

Arimino tenus munienda reliquas triumphalibus viris ex manubiali pecunia sternendas distribuit.

326. Suet. *Aug.* 37 (Augustus) *nova officia excogitavit: curam* [...] *viarum*; D.C. 54.8.4 *τότε δὲ αὐτός τε προστάτης τῶν περὶ τὴν Ῥώμην ὁδῶν* [...] *ὁδοποιοὺς αὐταῖς ἐκ τῶν ἐκστρατηγηκότων, ῥαβδούχοις δύο χρωμένους, προσέταξε.* According to Herzig (p. 628) this was valid for the immediate surroundings of Rome; *contra* Eck 1979, 26; van Binnebeke 43.
327. H.-C. Schneider 53. On the connection between consulates and the erection of milestones see Eck 1979, 26–7; Pekáry 75–7.
328. Fron. *Aq.* 101 *Itemque cum viarum curatores* [...] *publico fungantur ministerio, ut curatores aquarum iudiciis vacent privatis publicisque*; Eck 1992, 243.
329. Eck 1992, 237 and 244–5.
330. Eck 1979, 27–37, referring to inscriptions.
331. Cf. *CIL* IX 6072; Eck 1979, 35; Pekáry 93–4.
332. Eck 1979, 53–4; H.-C. Schneider 59; Pekáry 87; van Binnebeke 43–4.
333. Van Binnebeke 44; H.-C. Schneider 87.
334. Tac. *Ann.* 3.31.5 *Corbulo, plurima per Italiam itinera fraude mancipum et incuria magistratuum interrupta et impervia clamitando, executionem eius negotii libens suscepit*; D.C. 59.15.3–4 *Γναῖος Δομίτιος Κορβούλων βουλευτής* [...] *ἐζημίου*; Eck 1979, 57 and 66.
335. Eck 1979, 35.
336. Laurence 1999, 63 (ill.).
337. Contract workers: H.-C. Schneider 43; Pekáry 120 n. 151. Convicts and slaves: Suet. *Cal.* 27 *multos honesti ordinis deformatos* [...] *munitiones viarum* [...] *condemnavit*; Plin. *Ep. Tra.* 10.31 and 32, where there is talk of the use of slaves and convicts to public works. In the last letter, Trajan's answer, the emperor mentions road repair: *munitiones viarum et vicorum*; H.-C. Schneider 43. The army: *CIL* V 7989 (= *Supp. Ital.* 1316) *VIAM·QVOQVE | GEMINAM | A PORTA·VSQVE | AD·PONTEM | PER·TIRONES | IVVENTVT· NOVAE | ITALICAE·SVAE | DILECTVS POSTERIOR | LONGI TEMPORIS | LABE·CORRVPTAM | MVNIVIT·AC | RESTITVIT*; Pekáry 120 n. 152.
338. For more calculated examples see Pekáry 93–6; H.-C. Schneider 61.
339. For more examples of emperors who financed roads totally or partially see Eck 1979, 72–9.
340. Arist. *Ath.* 54.1 *ὁδοποιοὺς πέντε, οἷς προστέτακται δημοσίους ἐργάτας ἔχουσι τὰς ὁδοὺς ἐπισκευάζειν.*
341. For more information of *ὁδοποιοί, ἀστυνόμοι* and *ἀγρονόμοι* see also Pl. *Lg.* 763 C *ἕποιντο* [...] *τεταμένων*; Zimmermann 183.
342. Forbes 1934, 111–13. In *CIG* 4240 is talk of an *ἐπιμελητὴς ὁδῶν*; Radke 1903–78, 1475.
343. *Lex Julia Municipalis* 50–52 *Quominus aed(iles) et IIIIvir(ei) vieis in urbem purgandeis, IIvir(ei) vieis extra propiusve urbem Rom(am) passus M | purgandeis, queiquomque erunt, vias publicas purgandas curent eiusque rei potestatem habeant, | ita utei legibus pl(ebei)ve sc(itis) s(enatus)<ve> c(onsultis) oportet oportebit, eius h(ac) l(ege) n(ihil) r(ogatur)*. The citations of the *Lex Julia Municipalis* cited in this book are derived from *CIL* I^2 593; Frei-Stolba 25.
344. See p. 34 *supra*, incl. note 314.
345. Van Binnebeke 41–2 with references.
346. Van Binnebeke 53; she refers here to the *Lex Irnitana*, a municipal law of Irni (Spain).
347. *Lex Julia Municipalis* 20–55; van Binnebeke 58–9.
348. Van Binnebeke 61.
349. Paulus, *Sententiae* 5.6.2 *cuius rei sollicitudo ad viarum curatores pertinet, a quarum munitione nemo exceptus est*; Papin. *dig.* 43.10.1.3 *Ἐπισκευάζειν δὲ τὰς ὁδοὺς τὰς δημοσίας κατὰ τὴν ἑαυτοῦ οἰκίαν ἕκαστον καὶ τὰς ὑδρορρόας ἐκκαθαίρειν τὰς ἐκ τοῦ ὑπαιθρίου καὶ ἐπισκευάζειν οὕτως, ὡς ἂν μὴ κωλύῃ ἅμαξαν ἐπιέναι. ὅσοι δὲ μισθωσάμενοι οἰκοῦσιν, ἐὰν μὴ ἐπισκευάσῃ ὁ δεσπότης, αὐτοὶ ἐπισκευάσαντες*

ὑπολογιζέσθωσαν τὸ ἀνάλωμα κατὰ τὸν μισθόν. See van Binnebeke 63–97 for an elaborated explanation of the responsibilities of inhabitants concerning the maintenance of streets and roads.

350. Van Binnebeke 64, 72 and 95.
351. Papin. *dig.* 43.10.1.2 (damage to the road surface) Ἐπιμελείσθωσαν [. . .] καταλυέτωσαν; 43.10.1.5 (waste, skeletons and skins on the street) Μὴ ἐάτωσαν [. . .] μηδὲ κόπρον ἐκβάλλειν μηδὲ νεκρὰ μηδὲ δέρματα ῥίπτειν. For more information of street cleaning in Greek cities see Liebeschuetz 54–7.
352. Juv. 3.247 *pinguia crura luto*; Sen. *De ira* 3.35.5 *scabras lutosasque semitas spectant.*
353. Paul. *dig.* 18.6.13 (12) *Lectos emptos aedilis, cum in via publica positi essent, concidit.*
354. Papin. *dig.* 43.10.1.4 Ἐπιμελείσθωσαν δὲ καὶ ὅπως πρὸ τῶν ἐργαστηρίων μηδὲν προκείμενον ᾖ, πλὴν ἐὰν κναφεὺς ἱμάτια ψύγῃ ἢ τέκτων τροχοὺς ἔξω τιθῇ· τιθέσθωσαν δὲ καὶ οὗτοι, ὥστε μὴ κωλύειν ἅμαξαν βαδίζειν. Van Binnebeke 126; Frei-Stolba 36.
355. Mart. 7.61 (see p. 133); Matthews 26.
356. In Athens: Arist. *Ath.* 50.2 καὶ τὰς ὁδοὺς κωλύουσι (sc. οἱ ἀστυνόμοι) [. . .] τὰς θυρίδας εἰς τὴν ὁδὸν ἀνοίγειν; Zimmermann 183.
357. *Codex Iustinianus* 8.11(12).20 *Qui sine auctoritate divini rescripti ad iudicium tuae celsitudinis destinandi angiportus integros vel partes suis domibus incluserint seu porticus usurparint, procul dubio iura pristina sacratissimae reddere civitati iubemus: multa auri quinquaginta librarum non defutura, si quis posthac in similem audaciam prodire temptaverit.*
358. *Codex Theodosianus* 14.14 *Eos, qui in campo Martio casas seu tuguria conlocare temptaverint, sententia viri inl(ustris) p(rae)f(ecti) spoliatos omnibus facultatibus tradi in perpetuum exilio praecipimus*; van Binnebeke 141–2.
359. For an extended description of building in public spaces see van Binnebeke 133–68.
360. Stuart and de Grooth 41.
361. Willems 69.
362. Dahm 13.
363. Stuart and de Grooth 42–5; Riedel *passim.*
364. Riedel 8.
365. Stuart and de Grooth 42–3.
366. Riedel 8.
367. See the graduated scale on Figure 2.1.
368. For the terms 'double gate' and 'three-passage-gate' see chapter 3, pp. 92–4. For the situation of gates of Cologne see Riedel 38, Horn (1987) 465, Koepp 122 and Klinkenberg 183–97.
369. For example the stately home of *insula* 3 (Heimberg and Rieche 23 and 95).
370. Heimberg and Rieche 7, 90–5.
371. Etienne 1985, 151 and 304.
372. The distinction was not very strict here; there were also some stately homes on the south side.
373. D.C. 48.38.2 Ἀντώνιον [. . .] τὴν οἰκίαν αὐτοῦ τὴν πατρῴαν τὴν ἐν ταῖς Καρίναις κατεῖχε; Suet. *Tib.* 15 *e Carinis ac Pompeiana domo Esquilias in hortos Maecenatianos transmigravit.*
374. Friedländer 10.
375. See amongst others Mart. 7.61; Friedländer 21.

376. Already in early times: Liv. 6.38.8 *percitus ira Camillus lictores qui de medio plebem emoverent misit*; Juv. 3.128–9 *praetor lictorem inpellat et ire | praecipitem iubeat.*
377. A. Kolb 2000, 28–9.
378. Hudemann 9.
379. Liv. 42.1.7–11 [Postumius] *litteras* [...] *mos esset*; Hudemann 9.
380. A. Kolb 2000, 29–37.
381. Cic. *Att.* 5.16.
382. Hor. *S.* 1.5.45–6 *tectum | praebuit, et parochi, quae debent ligna salemque.* According to A. Kolb 2000 (p. 38) the *parochi* can mean the household of the house of Horace's friend L. Cocceius Nerva.
383. For an elaborate analysis of the abuses of the *cursus publicus* and its forerunners in general see the works of Hudemann and A. Kolb.
384. In some cases, it was possible to travel at night by *cursus publicus*, but it was an exception; André and Baslez 400.
385. A. Kolb 2000, 109–17, with clear table on p. 116.
386. *Historia Augusta, Pertinax* 1.6 *Dein praefectus cohortis in Syriam profectus Tito Aurelio imperatore a praeside Syriae, quod sine diplomatibus cursum usurpaverat, pedibus ab Antiochia ad legationem suam iter facere coactus est.*
387. A. Kolb 2000, 93–4.
388. Hudemann 42–3; *Codex Theodosianus* 12.12.9.1 *Illud etiam addimus, ut si integra dioecesis unum vel duos elegerit, quibus desideria cuncta committat, redae cursualis unius isdem tribuatur evectio.*
389. A. Kolb 2000, 89–92 with references; for bishops travelling at their own costs see n. 7. Of all British bishops in only three cases is it known that they used the *cursus publicus* (A. Kolb 2000, 90).
390. Ammianus Marcellinus 21.16.18 *excitavit discidia plurima* [...] *ut catervis antistitum iumentis publicis ultro citroque discurrentibus per synodos (quas appellant), dum ritum omnem ad suum trahere conatur arbitrium, rei vehiculariae succideret nervos.*
391. André and Baslez 216–20 with references.
392. One could expect of doctors that they had knowledge not only of medical science, but also of sciences such as astronomy and meteorology; such knowledge could be acquired only by travelling. A reference of this can be found in the *Corpus Hippocraticum*: Hp. *Aër.* 2 εἰ γὰρ ταῦτα [...] μεταβάλλουσιν τοῖσιν ἀνθρώποισιν; Brodersen 7–9 with references to Galen's travels.
393. Especially musicians and actors; André and Baslez 220–2 with references.
394. Cels. 5.27.3C *quod per quaedam medicamenta circulatores faciunt*; Petr. 68.6 *'et nunquam' inquit 'didicit, sed ego ad circulatores eum mittendo erudibam'.*
395. Stoffel 17; according to Hudemann (p. 116) a *statio* is the same as a *mansio*. Chevallier (1988, 67–8) enumerates several kinds of inns.
396. Heres 87; according to Hudemann (p. 116) the distance between *mutationes* in densely crowded regions is about 5 miles (2.5 hours to travel) and in less crowded regions 6 to 8 miles (4.5 hours to travel).
397. About the relation *mutationes/mansiones*, there are different points of view. Hudemann (p. 116) talks of six to eight *mutationes* between two *mansiones*; p. 121 quotes five to eight *mutationes* ('Ställe'); *contra* Stoffel (p. 17 n. 129), following the *Itinerarium Burdigalense*, amongst others 571.5 *mutationes CCXXX, mansiones CXII*; he talks of one *mansio* after two or three *mutationes*. According to Hudemann, in Late Antiquity there were more *mutationes* than in the preceding centuries (p. 116).
398. In Late Antiquity, a *mansio* had mule-drivers (*muliones*), donkey-drivers (ὀνελάται), horse assistants (*hippocomi*), veterinary surgeons (*mulomedici*) and coach-builders (*carpentarii*) available; the chief of a *mansio* was an inn keeper or, perhaps better, restaurant keeper (*manceps*); A. Kolb 2000, 197 with references especially to the *Codex Theodosianus.*

399. *CIL* IX 2689; for an illustration of the stone and complete Latin text see Heres 100.
400. For example Pompeii and Caerwent: Heres 95, 98 and 99.
401. Kleberg 16–17 incl. Tafel I; he gives still more inn names. For the situation of all inns found in Pompeii see the map on the last page of his book.
402. Pl. *Ps.* 558 *Ego devortor extra portam huc in tabernam tertiam*; Heres 99; Hudemann 124. For a concentration of inns at the city gates in Pompeii see Wallace-Hadrill 44 (Porta di Ercolano and Porta di Stabia).
403. Sen. *Ep.* 51.4 *Non tantum corpori, sed etiam moribus salubrem locum eligere debemus* [. . .] *inter tortores habitare nolim, sic ne inter popinas quidem*; Gross 56; Chevallier 1988, 77–8.
404. Hor. *S.* 1.5.7–8 *hic ego propter aquam, quod erat deterrima, ventri | indico bellum*; 1.5.14 *mali culices ranaeque palustres*; 1.5.19 *stertitque.*
405. Casson 127.
406. *AE* 1975, 429 *D(is) M(anibus) | permiss(u) manc(ipum) | iumentarior(um) | port(ae) Iou(iae)* [. . .] *Sosius | . . .ssem | . . .urno*; Riepl 197–8.
407. Etienne 1985, 191.
408. André and Baslez 413 with more places where *iumentarii* are found.
409. The scarce information about the actual distances caused failures in the annotation of distances in the guides; André and Baslez 379 with references.
410. Giebel 202.
411. Meijer 83; Giebel 179–203.
412. Suet. *Aug.* 47 *Nec est, ut opinor, provincia, excepta dum taxat Africa et Sardinia, quam non adierit*; Giebel 188.
413. André and Baslez 276–7.
414. Casson 130–31; André and Baslez 267–70 and 364 (map of all healing wells in Asia Minor).
415. Giebel 211.
416. On travelling with less comfort: Sen. *Ep.* 87.1–4, where Seneca travels by simple peasant's wagon, eating bread and figs; on the restless traveller see Sen. *De tranquillitate* 2.13–15; Giebel 211–14.
417. Crouwel 118.
418. Sizoo 125–42; Drijvers 174–9. A third *itinerarium*, the *Itinerarium Antonini*, dates *c.* 300 and derives from a road map like the Peutinger Map. For a detailed study of this *itinerarium* see Calzolari's article.
419. Giebel 218.
420. Hieronymus, *Epistulae* 58.4.4 on the crowd in Jerusalem: *urbe celerrima* [. . .] *in qua scorta, mimi, scurrae et omnia sunt, quae solent esse in ceteris urbibus*; Giebel 222–3.
421. The travel scheme and complete list of Theophanes' luggage and food comes to us by the papyri Gk 627–38, *Catalogue of the Greek Papyri in the John Rylands Library, Manchester*, nos 627–38. The shortened list is based on Casson 190–3. For the presence of bedding and material for cooking see also Gross 54–5.
422. Chapter 13 of André and Baslez (pp. 483–540) is spent completely on the traveller's discomfort.
423. Aristid. *Or.* 51.1–10; this work is also known as the 'Ιεροὶ λόγοι V. Heat: 51.2 *καῦμα ἤρετο λαμπρόν*; cold: 51.5 *ψῦχός* [. . .] *δεῖσθαι*; blowing sand: 51.9 *τότε δὲ τῆς τε ἄμμου πολλῆς ἅμα τῷ ἀνέμῳ φερομένης ἐπὶ τοῦ προσώπου*; closed inns: 51.4 *πάντα ἀπεκέκλειτο*; 51.5 *ἀνεῳγμένον μηδέν.*
424. Aristid. *Or.* 51.5 *παρακαλέσας* [. . .] *λοιπόν*; 14 *οἱ δ' ἀκόλουθοι ἀπειρήκεσαν.*
425. Also the fragments of the *Forma Urbis Romae* from the time of Septimius Severus do not mention street names, for example the fragments 165abd; for more information on this *Forma Urbis* see http://formaurbis.stanford.edu/index.php.
426. Etienne 1985, 299–300. The street names used nowadays date from modern times.
427. For coachmen rentals see p. 55.

428. Cross-breeding of a male ass and a mare; a cross-breeding of a stud and a she-ass is called *hinnus*: Var. *R.* 2.8.1 *Ex equa enim et asino fit mulus, contra ex equo et asina hinnus.*
429. Crouwel 115.
430. André and Baslez 409 with references to Varro and Columella.
431. Plin. *Nat.* 8.171 *Ex asino et equa mula gignitur mense XIII, animal viribus in labores eximium*; André and Baslez 409.
432. Von Stephan 44.
433. Wegener Sleeswyk 72.
434. Apul. *Fl.* 21 *quippe et illis, quibus curriculo confecta via opus est, adeo uti praeoptent pendere equo quam carpento sedere, propter molestias sarcinarum et pondera vehiculorum et moras orbium et salebras orbitarum - adde et lapidum globos et caudicum toros et camporum rivos et collium clivos–; hisce igitur moramentis omnibus qui volunt devitare ac vectorem sibimet equum deligunt diutinae fortitudinis, vivacis pernicitatis, id est et ferre validum et ire rapidum*; Chevallier 1988, 41–2.
435. Crouwel 115–17.
436. For an enumeration of all vehicles and their purposes see Lafaye, 'Vehiculum' 667.
437. Saglio, 'Carpentum' 926–7.
438. Isid. *Orig.* 20.12.3 *Carpentum pompaticum vehiculi genus.*
439. Apul. *Met.* 11.26 *Augusti portum celerrime <pervenio> ac dehinc carpento pervolavi.*
440. Saglio, 'Carruca' 928.
441. The *carruca* was popular with women: Scaev. *dig.* 34.2.13 *quaesitum est, an carrucha* [*sic*] *dormitoria cum mulis, cum semper uxor usa sit, ei debeatur.*
442. Isid. *Orig.* 20.12.3 *carruca undique contecta.*
443. Meijer 74.
444. Measured by myself.
445. Röring 50.
446. Excavations in Xanten (Capitol) have shown traces of wheel ruts of a wagon with an axle width of 1.10 m: Horn (2000) 270–1.
447. Röring 50; Grénier 1934, 375–7 for extended information on rut widths. For the 'nachleben' of standard gauge see also note 899 *infra.*
448. Tsujimura 61–2. Unfortunately, Tsujimura is not clear if these dimensions are inclusive or exclusive of the width of the rims. For dimensions in feet see Davies 69.
449. Saglio, 'Cisium' 1201.
450. *Codex Theodosianus* 6.29.2.2 *birotum vel veredum* [*postu*] *lare permissum est*; in 8.5.8.1 there is talk of a *birota.* Nonius Marcellus 86M.30 (p. 123) *cissium* [*sic*] *vehiculi biroti genus.*
451. Saglio, 'Rheda' 862–3.
452. Hudemann (pp. 145–6) states that the *rheda*, originally, was a wagon with two wheels; he also states that a *rheda* could be drawn by two or four horses; in the latter case see also the *Codex Theodosianus* 6.29.5.pr. *raedas, id [es]t quas quadrigas vel flagella appellant*; this is discussed by Hudemann (pp. 148–9).
453. See note 388 *supra.*
454. Cic. *Mil.* 29 *adversi raedarium occidunt.*
455. Isid. *Orig.* 20.12. For a still longer enumeration of wagons see Chevallier 1988, 35–8 and Marquardt 732–6.
456. Isid. *Orig.* 20.11; for the possibility of using sedans in the case of the smallest roads (*itinera*) see chapter 1, pp. 8–9.
457. *Edictum Diocletiani de pretiis* 15.33 [*raeda cum arcuatis rotis sine fer*] *ro *tribus milibus*]; 15.34 [*dormitorium cum vitutis rotis sin*] *e ferro *s*[*eptem milibus quingentis*]; 15.35 [*dormitorium cum arcuatis ro*] *tis sine f* [*erro*] **qu*[*attuor milibus*]; 15.37 [*carruca vituta sine fer*] *ro *septem milibus.* The character * must be read here as '*denarii*'.
458. *Edictum Diocletiani de pretiis* p. 255 (comm. Lauffer).

459. *Edictum Diocletiani de pretiis* 17.1a *redae in un<o> homin<e> per passus mille{s}* *II*; 17.2 *redae in solito onere per passus M* *XII*.
460. Cic. *Fam.* 2.19.2 *ut ad te statores meos et lictores cum litteris mitterem*; A. Kolb 2000, 20.
461. Liv. 22.11.5 *viatorem misit qui consuli nuntiaret*; A. Kolb 2000, 21.
462. Cic. *Att.* 5.20.7–9 *nunc publice litteras Romam mittere parabam* [. . .] *Nicanor in officio est et a me liberaliter tractatur. Quem, ut puto, Romam cum litteris publicis mittam, ut et diligentius perferantur, et idem ad me certa de te et a te referat.* Hudemann (p. 11) speaks of *cursores*, messengers from the slave and freedman ranks; according to Kolb, there is talk of it for the first time after the founding of the *cursus publicus*; A. Kolb 2000, 53.
463. Isid. *Orig.* 5.24.4 *Tabulae* [. . .] *epistolarum* [. . .] *portitores earum tabellarii vocabantur; nuntii*: Cic. *Att.* 11.24.4 *per litteras quidem aut per nuntium*. Sometimes a *nuntius* could be the same as a *tabellarius*: Liv. 42.37.6 *nuntio ipsius qui litteras attulerat.*
464. A. Kolb 2000, 21–2.
465. *CIL* XIV 2874 *CISIAREI · PRAENESTINEI · F · P · D · D|MAG · CVR· TOSENIANVS ·L · L · LICIN|M· POMPEIVS ·hELIOD|MINISTREI· NICEPHORVS · C · TALABARA · |NICEPHORVS · MITREI*; A. Kolb 2000, 22 n. 6.
466. Caes. *Civ.* 3.101 *nuntii de Caesaris victoria per dispositos equites essent allati*; A. Kolb 2000, 25–7.
467. Cic. *Fam.* 2.10.1 *Tu vide, quam ad me litterae non perferantur*; 2.9.1 *In his enim sum locis, quo et propter longinquitatem et propter latrocinia tardissime omnia perferuntur.*
468. About the name *cursus publicus* see A. Kolb 2000, 49–53.
469. Suet. *Aug.* 49.3 *Et quo celerius ac sub manum adnuntiari cognoscique posset, quid in provincia quaque gereretur, iuvenes primo modicis intervallis per militaris vias, dehinc vehicula disposuit. Commodius id visum est, ut qui a loco idem perferunt litteras, interrogari quoque, si quid res exigant, possint*; A. Kolb 2000, 54–5.
470. For a clear summary of the functioning of the message service compared with the *cursus publicus* see A. Kolb 2000, 298–9.
471. Eck 1979, 89–94; A. Kolb 2000, 158.
472. Eck 1979, 94–9; A. Kolb 2000, 152. For a complete overview of all *praefecti vehiculorum* of Italy see A. Kolb 2000, 162–5 with epigraphical references.
473. *CIL* X 6976 *L · BAEBIVS · L · F|GAL · IVNCINVS| . . . PRAEF · VEHICVLORVM*.
474. *AE* 1955, 225; Eck 1979, 90–2.
475. *CIL* VIII 12020 = *ILS* 1411 *Q · IVLIO MAXIMO | DEMETRIANO* [. . .] *PRAEF VEHICuLOR | PER BELGICAM ET DVAS | GERMAN.*
476. *CIL* VI 1641 *praef. VEHICVLORVM · PER · GALLIAS*.
477. *CIL* X 7585 *Q · GABINIO · BARBARO . . . PRAEF · VEHIC · PER · |FLAM*; Eck 1979, 100–3.
478. *Codex Theodosianus* 8.5.4.1 *super qua re* [. . .] *praefectos vehiculorum adque omnes, qui cursui publico praesunt, admoneri conveniet; CIL* VI 33714 *HIC POSITVS · EST · MAXIMVS . . . PRAEPOSITVS|DE VIA FLABINIA*; Eck 1979, 102–3 and 273.
479. Eck 1979, 103; he rejects the possibility that the *praefectus praetorio* was standing between them.
480. *ILS* 9018 *Honor[ati] | A. Vitellio Pap. Felici Honorato | eq. R., | f.a. at vehicula per Flaminiam. f.a.* means *fisci advocato*.
481. *CIL* VI 8543 *T · FLAVIO · AVG · LIB|SATVRNINO · TABVLA|RIO · A· VEHICVLIS*.
482. *IvEph* 855 [*T*] *Fl. Aug. lib* | [*P*]*ergamo* [. . .] *acceptorem ve-*| [*hic*]*ulorum*; for more subsidiary functions see Eck 1979, 108–10; A. Kolb 2000, 157–62.
483. Stat. *Silv.* 4.9.16–19 *priusquam | te Germanicus arbitrum sequenti | annonae dedit omniumque late | praefecit stationibus viarum.*
484. When this happened is not clear. Inscriptions show that this was the case in the third century; A. Kolb 2000, 159 with epigraphical references, amongst others *CIL* VI 31369 (in the year 226). According to Eck this was already in Nerva's time: Eck 1979, 98.

485. A. Kolb 2001, 102.
486. *SEG* 26 1392; A. Kolb 2000, 96.
487. H.-C. Schneider 91; for the discussion at this point see A. Kolb 2000, 68–9.
488. *Codex Theodosianus* 8.5.62 *velocis et clavularii cursus*; for other references see A. Kolb 2000, 52–3.
489. Bender 1978, 9; A. Kolb 2000, 96–7.
490. Hudemann 144–5.
491. *SEG* 16 754; A. Kolb 2000, 96.
492. *Codex Iustinianus* 12.50.22 *Cursum clavularem* [. . .] *in armorum tam confectione quam translatione; Codex Theodosianus* 8.5.33.1 *vestes militares*; A. Kolb 2001, 102 n. 51.
493. A. Kolb 2000 *passim*; for the transport of bishops see Optatus Milevitanus, *Appendix* VIII p. 212 *angarialem* [. . .] *ad Arelatensem portum* [. . .] *dedimus*; A. Kolb 2000, 88–9. For an explicit overview of wagons and animals of the *cursus publicus* see Stoffel 21–2.
494. *Codex Theodosianus* 8.5.47.pr. *carro sescentarum nec amplius addito eo*; Hudemann 152.
495. A Roman pound = 326 gr.
496. A. Kolb 2000, 96–7; Hudemann 149. See also Table 2.1 with references.
497. *Codex Theodosianus* 8.5.30 *Perspicue sanxeramus, ut in carpentis raedarum mensuram subditam nullus excederet, nemo amplius raedae quam mille pondo, angariae quam mille quingenta, veredo quam triginta auderet imponere* [. . .] *ut per loca, quae huiuscemodi observationis excubiis munienda sunt, sollicitos protectores diligentesque constituant, quo idem et mensuram vehic[u]lorum et vim onerum semper inspiciant nihilque fieri contra haec iura patiantur.*
498. *Codex Theodosianus* 8.5.17.1 *Illud sane, ut penitus enormium vehiculorum usus intercidat, sanciendum esse decernimus, ut, quisquis opificum ultra hanc quam perscribsimus normam vehiculum crediderit esse faciendum, non ambigat sibi, si liber sit, exilii poenam, si servus, metalli perpetua supplicia subeunda.*
499. *SEG* 19 476 [ταῖ]ς λιτουργίαις καὶ ἀ[ν] | [γαρεία]ις ὑπηρετοῦν[τ] / [*munera et*] *angarias pra*[*ebeant*]; A. Kolb 2000, 126; Stoffel 26. Lib. (50) criticises the ἀγγαρεία in Antioch, where peasants were forced to use their animals for the removal of rubble; Liebeschuetz 51–3. This ἀγγαρεία had nothing to do with the *cursus publicus.*
500. *CIL* III 7251 = *ILS* 214; A. Kolb 2000, 124.
501. Hudemann 17–18.
502. Hudemann 18; A. Kolb 2000, 140; Teitler 132.
503. Plin. *Ep.* 10.120; Tra. in Plin. *Ep.* 10.121.
504. Arcadius *dig.* 50.4.18.29 *Civilium munerum immunitas alicui concedatur, neque ab annona neque ab angariis neque a veredo* [. . .] *exceptis militibus et veteranis, excusari possunt.*
505. *Codex Theodosianus* 8.5.1 *qui enim explicaverit mansionem, si forte boves non habuerit, inmorari debet, donec fuerint exhibiti ab his, qui cursus publici curam gerunt, nec culturae terrae inservientes abstrahere*; Hudemann 30; Stoffel 37.
506. *Codex Theodosianus* 8.5.2 *Quoniam plerique nodosis et validissimus fustibus inter ipsa currendi primordia animalia publica cogunt quidquid virium habent absumere, placet, ut omnino nullus in agitando fuste utatur.*
507. *Codex Theodosianus* 8.5.5 *Iam dudum nostrae clementiae iussa exsistunt, ut rectoribus provinciarum evectionum faciendarum copia denegetur, quoniam cursui publico magna infertur pernicies, si haec licentia latius panderetur. Ideoque hoc ipsum repetimus quod ante placuerat ac iubemus eos auctoritatis tuae litteris commoneri, ut iussis parere festinent.*
508. *Codex Theodosianus* 8.5.15 *Mancipum cursus publici dispositio proconsulis forma teneatur, neque tamen sit cuiusquam tam insignis audacia, qui parangarias aut paraveredos in civitatibus ad canalem audeat commovere, quo minus marmora privatorum vehiculis provincialium transferantur, ne otiosis aedium cultibus provincialium patrimonia fortunaeque lacerentur.*
509. Sometimes one needed nearly twenty animals: Hudemann 34. Each year, a quarter of

the animals had to be replaced: *Codex Theodosianus* 8.5.34.pr. *veredorum pars quarta reparatur*; A. Kolb 2000, 130.

510. *CIL* V 8987 = *ILS* 755 *BREVIATIS · MVTATIONVM · SPA| TIIS.*

511. *Codex Theodosianus* 8.5.14 *Qui contra adnotationem manus nostrae plures quam evectio continebit veredos crediderit usurpandos, capitalem rem fecisse videbitur.*

512. *Codex Theodosianus* 8.5.16.pr. *In provincia Sardinia, in qua nulli paene discursus veredorum seu paraveredorum necessarii esse noscuntur, ne provincialium status subruatur, memoratum cursum penitus amputari oportere decernimus.*

513. A. Kolb 2000 *passim*; Hudemann *passim*.

514. A. Kolb 2000, 85 with references to the *Notitia Dignitatum*.

515. Bender 1978, 19.

516. Hudemann 121; Bender 1978, 10; Procop. *Arc.* 30.4 ἵπποι δὲ ἵσταντο ἐς τεσσαράκοντα ἐν σταθμῷ ἑκάστῳ.

517. A. Kolb 2000, 91 n. 3 and 130; Stoffel 22; *Codex Theodosianus* 8.5.35.pr. *A nullo umquam oppido aut frequenti civitate, mansione denique adque vico uno die ultra quinque veredorum numerus moveatur*; 8.5.40.1 *seni veredi, singulae etiam raedae per dies singulos dimittantur; Codex Iustinianus* 12.50.8 *deni veredi per dies singulos*. There is also a difference between summer and winter: *Codex Theodosianus* 8.5.8.2 *Octo mulae iungantur ad raedam aestivo videlicet tempore, hiemali decem.*

518. A. Kolb 2000, 219.

519. *Codex Theodosianus* 8.5.38 *Proficiscente vicario triginta asini, veredi decem tantummodo moveantur.*

520. *SEG* 26 1392; A. Kolb 2000, 219.

521. *SEG* 26 1392; A. Kolb 2001, 97.

522. In 430, there were 1174: *Codex Theodosianus* 6.27.23 *numero in mille centum septuaginta quattuor nominibus adsignato*. However, they did not all have a travelling function: A. Kolb 2000, 283. For a later extension see A. Kolb 2000, 283 n. 7.

523. Clauss 1078. Liebenam (p. 1592) quotes the number of 10,000 men in total.

524. Plb. 6.21.7–10 *παραγενομένων* [. . .] *ἴσους*.

525. Naerebout and Singor 265.

526. Clauss 1079.

527. Liv. 29.24.14 *supplevitque ita eas legiones ut singulae sena milia et ducenos pedites, trecenos haberent equites.*

528. Clauss 1079–80.

529. Clauss 1082.

530. Three legions are missing: the XVIIth, XVIIIth and XIXth. They became extinct during Varus' defeat in 9 AD; Liebenam 1605.

531. Clauss 1086.

532. *Notitia Dignitatum or.* 7.38–58; 5.48–66; Clauss 1101–2 and Liebenam 1623. Liebenam talks of forty-four *auxilia* and cohorts in the Danube region and 105 *auxilia* and cohorts in the other provinces in the east part of the Empire. The number of *cunei* according to Liebenam is forty-six. The numbers given by Clauss and Liebenam, referring to the *Notitia Dignitatum*, do not agree with those, according to Liebenam (p. 1624–5), mentioned in the *Notitia Dignitatum*.

533. Erdkamp 2001, 136. Clauss quotes the number of 524,000. MacMullen gives some scholars' estimations, including a million (Rostovtzeff). Most estimations, however, are about 500,000.

534. Agath. *Hist.* 5.13.7 *πέντε καὶ τεσσαράκοντα καὶ ἑξακοσίας χιλιάδας μαχίμων ἀνδρῶν τὴν ὅλην* [. . .] *δύναμιν*; MacMullen 460; Clauss 1099.

535. See note 527 *supra*.

536. Vegetius, *Epitoma rei militaris* 2.6.9 *legio* [. . .] *quae habet pedites sex milia centum, equites DCCXXX.*

537. Hyg. *De metatione castrorum* 1 *Papilio* [. . .] *tegit homines VIII. Plena centuria habet milites LXXX; erunt papiliones X. Contra* Fiebiger, who states that there were eight *contubernia* of ten soldiers. Archaeology, however, confirms Hyginus' information; Junkelmann 94.
538. Junkelmann 94.
539. These numbers are derived from Junkelmann (p. 49).
540. Junkelmann 64 and 93.
541. Junkelmann 64.
542. Junkelmann 50.
543. J. *BJ* 3.6.2 (115–126) Οὐεσπασιανὸς [. . .] συχνοί; 5.2.1 (47–50) Προϊόντι [. . .] οὐραγοί Arr. *Alan.* 1–11 Ἡγεῖσθαι [. . .] καθίστασθαι; Morin *passim.*
544. Tac. *Ann.* 1.51 *pars equitum et auxiliariae cohortes ducebant, mox prima legio, et mediis impedimentis sinistrum latus unetvicesimani, dextrum quintani clausere; vicesima legio terga firmavit, post ceteri sociorum*; Chevallier 1988, 250.
545. J. *BJ* 3.6.2 (124) ἡ φάλαγξ τὸ στῖφος εἰς ἓξ πλατύνασα, = 5.2.1 (48); Arr. *Alan.* 5 αὐτοὺς δὲ τοὺς πεζοὺς ἐπὶ τεσσάρων τεταγμένους ἰέναι; 6 ἐπὶ τεσσάρων ὡσαύτως; Morin 151 and 154.
546. Bender 1978, 19; Gross 54–5; von Stephan 27.
547. Ambrosius, *Expositio psalmi CXVIII* 5.2.1–3 *Miles qui ingreditur* [. . .] *requies ministretur*; Adams 37; Chevallier 1988, 251.
548. Ambrosius, *Expositio psalmi CXVIII* 5.2.3 *eliguntur civitates, in quibus triduum, quatriduum et plures interponantur dies, si aquis abundant, commerciis frequentantur*; Adams 37.
549. Bender 1978, 18–19.
550. *CIL* III 3385, mentioning fortifications along the Danube riverbank during the reign of Commodus: *BV*[*rgis*] | *A SOLO EXTRVCTIS ITEM PRAES* [*i*] | *DIS PER LOCA OPPORTVNA AD* | *CLANDESTINOS LATRVNCVLO* |*RVM · TRANSITVS OPPOSITIS*| *MVNIVIT*; *loca opportuna* must be seen here as crossings. For more references see Lécrivain 1469 n. 11.
551. Amongst others the soldier Apuleius in Nicomedia: Plin. *Ep.* 10.74 *Apuleius* [. . .] *miles, qui est in statione Nicomedensi*; for references to other cities see Lécrivain 1469.
552. *CIL* VIII 14603; here, it concerns a *praesidium* in Africa Proconsularis along the road; Lécrivain 1469 n. 9.
553. Suet. *Aug.* 32.1 *grassaturas dispositis per opportuna loca stationibus inhibuit; Tib.* 37 *In primis tuendae pacis a grassaturis ac latrociniis* [. . .] *licentia curam habuit. Stationes militum per Italiam solito frequentiores disposuit.*
554. Ulp. *dig.* 11.4.1.2 *milites stationarios dominum adiuvare debere inquirendis fugitivis.*
555. Nelis-Clément 247–51; chapter 3, p. 89.
556. Nelis-Clément 75.
557. See note 554 *supra*; Ulp. *dig.* 1.12.1.12 (about the *praefectus urbi*): *sane debet etiam dispositos milites stationarios habere ad tuendam popularium quietem et ad referendum sibi quid ubi agatur.*
558. *AE* 1937, 250 Ζωτικος Ασκλ[ηπιω?] | κηρυξ ιππευς | σινγυλαριος στα[τιωνα] | ριος ευχ[ην]. For more epigraphical references to *stationarii* see Nelis-Clément 104–5 n. 112.
559. That *beneficiarii* had a higher position than *stationarii* can be proved by the difference of payment: 2 drachms and 2 obols for a *stationarius* and 4 drachms for a *beneficiarius*; Nelis-Clément 248–9 (nn. 148 and 150) with references to the papyri *P. Oxyr.* 1650 and 1650a.
560. Nelis-Clément 161–75; 511–25 (ill.).
561. Nelis-Clément 145.

562. Nelis-Clément *passim*; cities as amongst others Sirmium and Cologne: Riedel 20; Bender 1989, 132.
563. Plin. *Ep. Tra.* 10.77 *Providentissime, domine, fecisti, quod praecepisti Calpurnio Macro, clarissimo viro, ut legionarium centurionem Byzantium mitteret.* Nelis-Clément writes, however, *regionarium.* Trajan's answer: Tra. Plin. *Ep.* 10.78 *Ea condicio est civitatis Byzantiorum confluente undique in eam commeantium turba, ut secundum consuetudinem praecedentium temporum honoribus eius praesidio centurionis legionarii consulendum habuerimus.*
564. Nelis-Clément 176–7.
565. Nelis-Clément 75, especially n. 91 and 311 (C. Julius Valens).
566. On the discussion whether the frontier security of Hadrian's Wall was executed by provincial auxiliary troops or *beneficiarii* and/or *stationarii* see Webster 75 n. 2. For the security of the *limes* in south-west Germany, there was, in any case, besides normal soldiers, also talk of *beneficiarii*; Nelis-Clément *passim.*
567. Plin. *Ep.* 6.25; *Cambridge Ancient History* XI, 387.
568. *CIL* VIII 2728 + 18122 = *ILS* 5795 *PROFECTVS · SVM · E · IN|TER · VIAS· LATTRONES ·SVM|PASSVS · NVDVS· SAVCIVS ·E|VASI CVM · MEIS*; Grünewald 31–5, in which he casts doubt on the effectiveness of road security.
569. André and Baslez 129–30.
570. Also Epictetus gave the advice not to travel alone: Epict. *Diatribae* 4.1.91–95 Οὕτως [. . .] θνητός ἐστιν.
571. They also could use the unpaved strips of the highways 28–9. Besides that, they could also use the so-called *actus sine itinere*, see *supra* (p. 9).
572. Naerebout and Singor 366.
573. Heichelheim 879 (50–70 million); von Stephan 56 (90 million).
574. See note 851 *infra.*
575. Van Tilburg 2002, 192; Naerebout and Singor 364; Stark (pp. 131–2) gives a list of twenty-two cities with their numbers of inhabitants, where Constantinople is missing; Rome has no more than 650,000 inhabitants.
576. Yeo 221.
577. In this book, there is little attention to exotic trade with other countries.
578. Morley *passim.*
579. According to Garnsey (1988, 190–1 and 214) some production of corn in Italy was still maintained.
580. Morley 55–7.
581. Suet. *Nero* 16.1 *Destinarat etiam Ostia* [. . .] *atque inde fossa mare veteri urbi inducere*; Morley 66.
582. Var. *R.* 2.6.5 *Greges fiunt fere mercatorum, ut eorum qui e Brundisino aut Apulia asellis dossuariis comportant ad mare oleum aut vinum itemque frumentum aut quid aliut*; Morley 68.
583. Morley 68; White 128–9.
584. Cato *Agr.* 1.3 *aut mare aut amnis, qua naves ambulant*; Yeo 230.
585. Cic. *Phil.* 12.24 *Hisce ego me viis committam* [. . .] *nuper in suburbium, ut eodem die reverterer, ire non sum ausus?*
586. Cato *Agr.* 1.3 *via bona celerisque* [sc. *siet*]; Var. *R.* 1.16.1 *viae aut fluvii, qua portetur, aut non sunt aut idonei non sunt*; Col. 1.3.3 *Multum conferre agris iter commodum: primum, quod est maximum, ipsam praesentiam domini, qui libentius commeaturus sit, si vexationem viae non reformidet.*
587. Col. 1.3.3–4 *Deinde ad invehenda et exportanda utensilia, quae res frugibus conditis auget pretium, et minuit impensas rerum invectarum, quia minoris apportentur eo, quo facili nisu perveniatur; nec non nihil esse etiam parvo vehi, si conductis iumentis iter facias, quod magis expedit quam tueri propria.*

588. Apples: Hor. *S.* 2.4.70 *pomis Tiburtia*; figs: Plin. *Nat.* 15.69–70 *ficorum* [. . .] *Tiburtina*; mulberries: according to Morley 107; his reference to Plin. *Nat.* 15.97 talks of *mora Ostiensia et Tusculana.*
589. Morley 107; the numbers agree with Fig. 2.7.
590. Jul. *Mis.* 369 A ἔδοξέ μοι πέμπειν εἰς Χαλκίδα καὶ ʽΙεράν πόλιν [. . .] μυριάδας; Garnsey 1988, 23; Finley 33–4.
591. For the aspects of the various animals see chapter 2, pp. 51–2.
592. White 129. Erdkamp (p. 72 n. 91) enumerates numbers of pack and draught animals, roughly corresponding with those of White. Yeo quotes 250 pounds as the weight for a pack ass (p. 225).
593. White 133 and 208 (appendix); Chevallier 1988, 37.
594. Erdkamp 73 n. 92.
595. Var. *R.* 1.20.3 (*iuvenci*) *Quos ad vecturas, item instituendum ut inania primum ducant plaustra et, si possis, per vicum aut oppidum*; 2.8.5 *Hisce* (i.e. *muli*) *enim binis coniunctis omnia vehicula in viis ducuntur*; Raepsaet 1996–, 1100.
596. According to Pliny the Elder, the four-wheeled cargo wagon was probably invented by the Phrygians; Plin. *Nat.* 7.56 *vehiculum cum quattuor rotis Phryges.*
597. White 134–5. Raepsaet (1996–, 1102–3) mentions that both types of wagons were in use, with or without revolving front axles.
598. Yeo 225.
599. Raepsaet 1996–, 1101.
600. Raepsaet (1996–, 1101) places the *plaustrum* among the Celtic wagon types, but according to Lafaye (p. 505), it was an Italian wagon. The evidence for this could be that the word *plaustrum* is derived from *plaudere*, 'rattle'; cf. A. Walde and J.B. Hofmann, *Lateinisches etymologisches Wörterbuch*, Heidelberg 1930–56, 320 [Vol. II].
601. Verg. *G.* 2.206 *tardis decedere plaustra iuvencis*; see note 595; Lafaye, 'Plaustrum' 505.
602. Lafaye, 'Plaustrum' 506. Chevallier also mentions a so-called *plaustrum minus*, with two wheels (1988, 36).
603. *Lex Julia Municipalis* 62–64 *Quibus diebus virgines Vestales, regem sacrorum, flamines plostreis in urbe sacrorum publicorum p(opuli) R(omani) caussa | vehi oportebit, quaeque plostra triumphi caussa, quodie quisque triumphabit, ducei oportebit, quaeque | plostra ludorum.*
604. Saglio, 'Carrus' 928–9; Erdkamp 72 n. 91.
605. Lafaye, 'Vehiculum' 667. Also Chevallier (1988, 36–8) summarises all types of wagons. Besides goods transport, these wagons were also used for travellers.
606. Junckelmann 64.
607. Wallace-Hadrill 44–9; see also chapter 4, pp. 140–1.
608. Etienne 1987, 235–49.
609. Garnsey 1983, 62.
610. Virlouvet 177. For the measure of a *modius*, varying between 6.5 and 7.5 kg, see n. 7.
611. Ph. *Legatio ad Gaium* 158 οὐ μὴν ἀλλὰ κἀν ταῖς μηνιαίοις τῆς πατρίδος διανομαῖς, ἀργύριον ἢ σῖτον ἐν μέρει παντὸς τοῦ δήμου λαμβάνοντος, οὐδέποτε τούς ʼΙουδαιους ἠλάττωσε τῆς χάριτος, ἀλλʼ εἰ καὶ συνέβη τῆς ἱερᾶς ἑβδόμης ἐνεστώσης γενέσθαι τὴν διανομήν [. . .] προσετέτακτο τοῖς διανέμουσι ταμιεύειν τοῖς ʼΙουδαίοις εἰς τὴν ὑστεραίαν τὴν κοινήν φιλανθρωπίαν.
612. *Lex Julia Municipalis* 15–16 *quom frumentum populo dabitur, ibei ubei frumen|tum populo dabitur cottidie maiorem partem diei propositum habeto. Cottidie* can connect to *quom frumentum populo dabitur*; Virlouvet 179–80.
613. Suet. *Jul.* 41 (Caesar) *ex viginti trecentisque milibus accipientium frumentum e publico ad centum quinquaginta retraxit.*
614. Virlouvet 182.

615. Pouilloux 128, 58–60 ποιείσϑωσαν δὲ τὴν μέτρησιν ἀπὸ νουμηνίας ἕως δεκάτης, τοῖς δὲ ἀποδημοῦσιν ἐὰν ἔλϑωσιν ἕως τριακάδος.
616. Legras 58; Virlouvet 187–9; van Heck 316 (182.2 n. 1).
617. Lanciani 28; Lugli and Gismondi.
618. Morley 55.
619. For collecting taxes at the gates of the Aurelian Wall see chapter 3, pp. 87–8.
620. Bender 1978, 18–19.
621. *Historia Augusta, Severus Alexander* 45.2 *itinerum autem dies publice proponebantur, ita ut edictum penderet ante menses duos, in quo scriptum esset, 'Illa die, illa hora ab urbe sum exiturus et, si di voluerint, in prima mansione mansurus', deinde per ordinem mansiones, deinde stativae, deinde ubi annona esset accipienda.*
622. Suet. *Tib.* 38 *vehiculis comprehensis, commeatibus per municipia et colonias dispositis*; Halfmann 75.
623. For the army as road-user see chapter 2, pp. 63–8.
624. Adams 38.
625. Giebel 191.
626. Halfmann 76. For Plotius Grypus' function see p. 58; A. Kolb 2000, 159.
627. *ILS* 1140 <Tib. Cl. Candido> . . . *praeposito copiarum expediti|onis Germanicae secundae*; Halfmann 77.
628. Halfmann 78. For the role of the *cursus publicus* in Late Antiquity see also chapter 2, pp. 60–1.
629. Suet. *Aug.* 82.1 *Itinera lectica* [. . .] *procederet; Tib.* 60 *In quodam itinere lectica, qua vehebatur,* [. . .] *exploratorem viae, primarum cohortium centurionem.*
630. Hdn. 4.7.6 τὰ πλεῖστά τε αὐτοῖς συνώδευε περιπατῶν, σπανίως ὀχήματος ἢ ἵππου ἐπιβαίνων, τά τε ὅπλα βαστάςων ἑαυτῷ.
631. Castritius 365 n. 1 with references to amongst others Eutropius 9.24 *per aliquot passuum milia purpuratus tradatur ad vehiculum cucurrisse.*
632. Halfmann 89.
633. Gal. *De praecogn.* 9 (14.648–651 K.); Halfmann 108–9.
634. Cic. *Att.* 13.52 *villa ita completa a militibus est, ut vix triclinium, ubi cenaturus ipse Caesar esset, vacaret, quippe hominum* C|Ɔ C|Ɔ (2000 men).
635. *P. Oxyr.* 3602–3605; Halfmann 84 and 110.
636. Suet. *Cal.* 39 *vehiculis et pistrinensibus iumentis, adeo ut et panis Romae saepe deficeret; Nero* 30.3 *Numquam minus mille carrucis fecisse iter traditur, soleis mularum argenteis*; Giebel 212 (Unfortunately, I could not find the original citation of Poppaea's she-asses).
637. *Historia Augusta, Heliogabalus* 31.5 *imperator vero etiam sescenta vehicula dicitur duxisse, adserens decem milibus camelorum Persarum regem iter facere et Neronem quingentis carrucis iter inisse.*
638. Lib. *Or.* 11.178 ὕδωρ [. . .] πινόμενον; a reference to Xerxes' army, which also required a lot of water: Hdt. 7.21.1 κοῖον δὲ πινόμενόν μιν ὕδωρ οὐκ ἐπέλιπε πλὴν τῶν μεγάλων ποταμῶν; Halfmann 110 and 133.
639. Halfmann 138–40.
640. Plin. *Pan.* 20; Giebel 190.
641. For more information on triumphal processions see the PhD thesis of H.S. Versnel, *Triumphus: An inquiry into the origin, development and meaning of the Roman triumph*, Leiden 1970; F. Coarelli, *Il foro Boario: dalle origini alla fine della repubblica*, Rome 1988.
642. Reinach 536 with references.
643. Scullard 1974, 24–31; Isid. *Orig.* 14.5.12 [Mauretania] *Olim etiam et elephantis plena fuit, quos sola nunc India parturit.*

644. D.C. 60.21.2 παρασκευή γε ἐπὶ τῇ στρατείᾳ πολλὴ τῶν τε ἄλλων καὶ ἐλεφάντων προσυνείλεκτο.
645. Juv. 12.102–107 *quatenus hic non sunt nec venales elephanti, | nec Latio aut usquam sub nostro sidere talis | belua concipitur, sed furva gente petita | arboribus Rutulis et Turni pascitur agro, | Caesaris armentum nulli servire paratum | privato.*
646. *Historia Augusta, Gordiani tres* 33.1 *Fuerunt sub Gordiano Romae elefanti triginta et duo, quorum ipse duodecim miserat*; 27.9 *His in senatu lectis quadrigae elefantorum Gordiano decretae sunt, utpote qui Persas vicisset, ut triumpho Persico triumpharet*; Scullard 1974, 201–2.
647. Reinach 542; Plu. *Pomp.* 14.4 ὁ Πομπήϊος ἐπεχείρησεν ἐλεφάντων ἅρματι τεττάρων ἐπιβὰς εἰσελαύνειν· [...] ἀλλὰ τῆς πύλης στενωτέρας οὔσης, ἀπέστη καὶ μετῆλθεν ἐπὶ τοὺς ἵππους.
648. *Historia Augusta, Heliogabalus* 23.1 *Fertur* [...] *elefantorum quattuor quadrigas in Vaticano agitasse dirutis sepulchris quae obsistebant.*
649. *Historia Augusta, Aurelianus* 33.3–4 *Capitolium Aurelianus invectus est* [...]. *Praecesserunt elephanti viginti.*
650. Plin. *Nat.* 5.10 *crocodilis et hippopotamis refertum.*
651. Plin. *Nat.* 8.96 *M. Scaurus* [...] *hippopotamus in quadam medendi parte etiam magister existit.* For the spelling *hippopotamus*, I follow the Loeb text spelling. The Teubner edition gives the spelling *hippopotamius.*
652. *Historia Augusta, Heliogabalus* 28.3 *Habuit et hippopotamos; Gordiani tres* 33.1 *Fuerunt sub Gordiano Romae* [...] *hippopotami sex*; Hünemörder, 'Nilpferd' 946–7.
653. Plin. *Nat.* 11.128 *asino Indico qui uno armatus est cornu*; 255 *unicorne asinus tantum Indicus.*
654. Plin. *Nat.* 8.71 *Isdem ludis et rhinoceros unius in nare cornus, qualis saepe visus.*
655. Hünemörder, 'Nashorn' 921–2.
656. Mart. 1.3.5–6 *invenesque senesque | et pueri nasum rhinocerotis habent*; *Historia Augusta, Gordiani tres* 33.1 *Fuerunt sub Gordiano Romae* [...] *rhinoceros unus.*
657. Plin. *Nat.* 8.69 *camelopardalis, dictatoris Caesaris circensibus ludis primum visa Romae*; D.C. 43.23.1 περὶ δὲ δὴ τῆς καμηλοπαρδάλιδος ὠνομασμένης ἐρῶ, ὅτι τότε πρῶτον ἔς τε τὴν Ῥώμην ὑπ' αὐτοῦ ἐσήχθη καὶ πᾶσιν ἐπεδείχθη. An extended description of the physiognomy of the giraffe follows hereafter.
658. Ath. *Diphosophistae* 5.201C.
659. Hor. *Ep.* 2.1.195 *diversum confusa genus panthera camelo.* The Latin name of giraffe is *camelopardalis.* For the hypothesis that here we have to do with a giraffe see also the commentary of A. Kiessling and R. Heinze, *Q. Horatius Flaccus: Briefe*, Berlin 1961^7, 233.
660. *Historia Augusta, Gordiani tres* 33.1 *camelopardali decem* [...] *Philippus ludis saecularibus vel dedit vel occidit*; Hünemörder, 'Giraffe' 1075.
661. *Codex Theodosianus* 8.6.2 *qui animalia* [...] *prosequuntur*; A. Kolb 2000, 97–8.
662. A. Kolb 2000, 40.
663. These and the following dimensions are given by T. Dorresteijn of Diergaarde Blijdorp, Rotterdam.
664. See chapter 1, p. 31.
665. See Table 3.1, pp. 108–9.
666. Raepsaet 1984, 103.
667. *IG* II^2 II –2, 1673, 31 ἄξονας τοὺς ἐν τῷ ἐνιαυτῷ [κ]αταγέντας: ΔΓII (axles); 33 ἧλοι, ὧν προσεδέησεν: ΓΔΓIII (pins); Raepsaet 1984, 105 and 119; van Vliet 20.
668. A reconstruction is depicted by Korres (p. 37). With this vehicle, a monolithic capital could be transported. Significant are the broad wheels and the revolving front gearwheel.
669. MC (Mons Claudianus) Inv. No. O.7334; Peacock and Maxfield 262.
670. Peacock and Maxfield 263.

671. Ammianus Marcellinus 17.4.12–14 *Constantium* [. . .] *per Ostiensem portam piscinamque publicam, Circo illatus est Maximo.*
672. DeLaine 99.
673. DeLaine 98.
674. Peacock and Maxfield 263.
675. Van Vliet 30–1.
676. The French were confronted with the same problem in 1832 with their transport of the obelisk from Luxor; DeLaine 99 n. 73.
677. Liv. 2.9.6 *portoriisque et tributo plebes liberata*; de Laet 45. Toll in Rome: de Laet 113; Cagnat 587; Palmer *passim*.
678. De Laet 52 n. 1.
679. De Laet 114-15 and 346. It is not clear to me whether these *stationes* were also places along roads.
680. Cic. *Font.* 19 *Cognoscite* [. . .] *exegisse*; de Laet 83.
681. Cic. *Font.* 19 *Cognoscite nunc de crimine vinario, quod illi invidiosissimum et maximum esse voluerunt.*
682. De Laet 99.
683. Hdn. 2.4.7 τέλη τε πάντα <τὰ> πρότερον ἐπὶ τῆς τυραννίδος ἐς εὐπορίαν χρημάτων ἐπινοηθέντα ἐπί τε ὄχθαις ποταμῶν καὶ λιμέσι πόλεων ἔν τε ὁδῶν πορείαις καταλύσας ἐς τὸ ἀρχαῖον καὶ ἐλεύθερον ἀφῆκεν; de Laet 119-22; Cagnat 587.
684. De Laet *passim*, amongst others Carte 1 (p. 126); Cagnat 587–90.
685. De Laet 368; Masquelez and Saglio, 'Cataracta' 967.
686. De Laet 164–6; Johnston 26.
687. Cartillieri 23 and 75; Walser *passim*.
688. Suet. *Cal.* 40 *Vectigalia* [. . .] *essent.*
689. Plin. *Nat.* 19.56 (Rackham) *herba* [. . .] *Nullum quam macelli vectigal maius fuit Romae clamore plebis incusantis apud omnes principes, donec remissum est portorium mercis huius.*
690. *CIL* VI 1016a *IMP CAESAR M AVRELIVS | ANTONINVS·AVG* [. . .] *ANSARII (sic) | PROMERCALIVM·SECVNDVM | VETEREM·LEGEM·SEMEL·DVM | TAXAT·EXIGVNDO*; 1016b *IMP·CAESAR·M·AVRELIVS | ANTONINVS·AVG* [. . .] *ANSARII | PROMERCALIVM·SECVNDVM | VETEREM· LEGEM·SEMEL·DVM | TAXAT·EXI GVNDO*; 1016c *IMP CAESAR M AVRELIVS | ANTONINVS·AVG* [. . .] *ANSARII | PROMERCALIVM·SECVNDVM | VETEREM·LEGEM·SEMEL· DVM | TAXAT·EXI GVNDO; CIL* VI 8594 *QVIDQVID VSVARIVM INVEHITVR | ANSARIVM NON DEBET*; de Laet 348–9. These inscriptions are found along Via Flaminia and Via Salaria; *CIL* VI 8594 along the Tiber. These are the only places where the word *ansarium* is used.
691. Palmer 1980, 219.
692. Palmer 1980, 219–20.
693. Palmer 1980, 223.
694. Palmer 1980, 220–4.
695. Cagnat 592.
696. Ulp. *dig.* 49.14.6.1 *Quodcumque privilegii fisco competit, hoc idem et Caesaris ratio et Augustae habere solet.*
697. *Codex Iustinianus* 4.61.8 *A legatis gentium devotarum ex his* [. . .] *speciebus* [. . .] *quas vero ex Romano solo, quae sunt tamen lege concessae, ad propria deferunt, has habeant a praestatione immunes ac liberas.*
698. Symmachus, *Epistulae* 5.62 *Quaestores ordinis nostri numquam ferarum suarum portorium contulerunt: grave enim maioribus visum est ut tolerantibus onera senatoriae dignitatis sumptus immodicus adderetur.*

699. *Codex Theodosianus* 13.5.23 *navicularios a vectigali praestatione inmunes esse praecipimus.*
700. Paul. *dig.* 39.4.4.1 *De rebus, quas in usus advehendas sibi mandant praesides, divus Hadrianus praesidibus scripsit, ut, quotiens quis in usus aut eorum, qui provinciis exercitibusve praesunt, aut procuratorum suorum usus sui causa mittet quendam empturum, significet libello manu sua subscripto eumque ad publicanum mittat, ut, si quid amplius quam mandatum est transferet, id munificum sit.* Palace functionaries: *Codex Theodosianus* 11.12.3 *hisve, qui in palatio nostro degunt*; de Laet 478–9.
701. Tac. *Ann.* 13.51 *edixit princeps* [. . .] *militibus immunitas servaretur, nisi in iis, quae veno exercerent*; de Laet 432–5 and 478–9 with references.
702. Quint. *Decl.* 359 *Praeter instrumenta itineris omnes res quadragesimam publicano debeant. Publicano scrutari liceat. Quod quis professus non fuerit, perdat. Matronam ne liceat attingere*; de Laet 425-7. According to Cagnat (pp. 593–4) in the case of camel caravans even these *instrumenta itineris* were not completely tax-free.
703. *Codex Iustinianus* 4.61.5 *Universi provinciales pro his rebus, quas ad usum proprium vel ad fiscum inferunt vel exercendi ruris gratia revehunt, nullum vectigal a stationariis exigantur.* This is an edict of Constantine. In legal texts, the word *vectigal* is used, not *portorium.*
704. See p. 87.
705. Paul. *dig.* 39.4.9.7 *Res exercitui paratas praestationi vectigalium subici non placuit*; 39.4.9.8 *Fiscus ab omnium vectigalium praestationibus immunis est*; Cagnat 592.
706. *Codex Iustinianus* 3.44.15 Ἐν οὐδενὶ τόπῳ οὐδὲν τέλος παρ᾽ οὐδενὸς προσώπου ὑπὲρ τῶν διακομιζομένων σωμάτων παρέχεται; Cagnat 593.
707. *CIL* III 5122 *LIBERO | PATRI | SACR | ABASCANTVS | ANTONI·RVFI·S| SCR·V · T · V · S · L· M; CIL* III 14354[27] *INVICT·MITHRAE E TRANSITV·DEI | THEODORVS·P(ublici)·P(ortorii) | SCRVT(ator)· STAT(ionis)·POET(oviniensis) | EX·VISV.*
708. De Laet 141-2. On p. 153 he mentions detachments at the toll point of Zürich.
709. *Matronae* were not frisked (see note 702 *supra*), but were nevertheless forced to surrender certain goods; Quint. *Decl.* 359; Casson 290–1; Friedländer 303–4 (pearls).
710. Plu. *Moralia* 518E καὶ γὰρ τοὺς τελώνας [. . .] ποιοῦντες; Casson 290–1.
711. Palmer 222–3.
712. Var. *L.* 5.143 *Oppida* [. . .] *murum*; Coarelli 1997, 91. See also note 939 *infra.*
713. Serv. 5.755 *conditores* [. . .] *aratrum suspendentes circa loca portarum.*
714. Var. *L.* 5.143 *orbis* [. . .] *qui quod erat post murum, postmoerium dictum.*
715. Liv. 1.44.4 *Pomerium* [. . .] *est autem magis circamoerium.*
716. Vitr. 1.5.2 *uti portarum itinera non sint directa sed scaeva.*
717. Paul. *Fest.* p. 258M (p. 310 Lindsay) *Quadrata Roma in Palatio*; D.H. *Antiquitates Romanae* 2.65.3 τῆς τετραγώνου καλουμένης Ῥώμης ἣ ἐκεῖνος ἐτείχισεν; cf. van Heck 460; Giovagnoli 21 and 23. The word *quadrata* is a reference to the square plots of the inhabitants, cf. Var. *L.* 5.34 *ager* [. . .] *in quadratum actum*; 5.35 *Iugerum dictum iunctis duobus actibus quadratis*; Palmer 26–34.
718. Giovagnoli 23–4.
719. Plin. *Nat.* 3.66 *Urbem III portas habentem Romulus reliquit, ut plurimas tradentibus credamus, IV. Contra* some authors mention only one gate at Palatine Hill: amongst others Liv. 1.12.3 *Ad veterem portam Palati*; Platner and Ashby 376.
720. Var. *L.* 5.164–165 *portas* [. . .] *Mucionis* [. . .] *alteram Romanulam* [. . .] *Tertia est Ianualis* [. . .] *ut sit aperta semper, nisi cum bellum sit nusquam*; Giovagnoli 23–4. For the questionable situation of the gates see Holland, *passim*; she mentions Scalae Caci as third gate (p. 257 n. 51). For the opening of Porta Janualis see Macr. 1.9.17 *Romani portam* [. . .] *Ianualis vocata est, claudere festinarent.*

721. Fougères 584; Platner and Ashby 376. *Contra* Scullard 1967, amongst others 105 (Veii) and 121 (Vulci). A reference to the gods is the quotation of Serv. *A.* 1.422 *prudentes Etruscae disciplinae aiunt apud conditores Etruscarum urbium non putatas iustas urbes fuisse in quibus non tres portae essent dedicatae et tot viae et tot templa, Iovis Iunonis Minervae*; Thulin 1906–9, 3: 30.
722. Scullard 1967, 147 and illustration 68. For other Republican gates see Brands *passim*.
723. Liv. 1.44.3 *Aggere et fossis et muro circumdat urbem; ita pomerium profert.*
724. Giovagnoli 28.
725. Giovagnoli 29.
726. Stambaugh 20.
727. Palmer 308-9; Lugli and Gismondi. Palmer describes that *portae* also means 'customs stations'; Palmer 1980, 217 and 223.
728. Giovagnoli 29–38.
729. Plin. *Nat.* 3.66 *portas, quae sunt hodie numero XXXVII*; Coarelli 1997, 89.
730. Säflund xviii–xix; 191–2; 196–1; 220. See also Lugli and Gismondi.
731. Rackham's edition mentions in a note that the Twelve Gates are 'double gates'.
732. Detlefsen's edition uses lower case.
733. The Circus Maximus has twelve *carceres* (vaulted rooms) on the short flat side; cf. Cassiodorus, *Variae* 3.51.4 *Bis sena quippe ostia ad duodecim signa posuerunt.* Platner and Ashby (p. 119) assume that this is the origin of the name *duodecim portae* and I agree with this.
734. D.H. *Antiquitates Romanae* 4.13.5 εἰ δὲ τῷ τείχει, δυσευρέτῳ μὲν ὄντι διὰ τὰς περιλαμβανούσας αὐτὸ πολλαχόθεν οἰκήσεις; Holland 249 n. 19; Säflund 190.
735. Lugli and Gismondi.
736. Plin. *Nat.* 3.66 *praetereanturque ex veteribus VII quae esse desierunt.* Coarelli (1997, 89 n. 2) enumerates the names of the gates which had disappeared.
737. Säflund 65; Platner and Ashby 419.
738. See *supra*.
739. Schultze 284.
740. Etienne 1985, 99.
741. Brands 197–8; ill. 222.
742. Platner and Ashby 418.
743. D.H. *Antiquitates Romanae* 1.32.2 ταῖς Καρμεντίσι πύλαις.
744. Liv. 2.49.8., see also Ov. *Fast.* 2.201 *Carmentis portae dextro est via proxima Iano.*
745. Holland 249 and 250.
746. Holland 250 n. 21 (Kähler's quotation concerns triumphal arches, not city gates); Säflund 194-5. Säflund states that, maybe, Porta Carmentalis became a double gate after the Second Punic War.
747. Holland 249 n. 19; Säflund 195.
748. Etienne 1985, 95–7.
749. Porta di Stabia is discussed on pp. 110–1.
750. In many former publications a Porta di Capua was mentioned, but based on misunderstandings; now it is definitely stated that there was no gate between Porta del Vesuvio and Porta di Nola but Torre XI. For extended information about the research and misunderstandings see Sakai's extract.
751. Etienne 1985, 96.
752. The date of Porta di Ercolano is doubtful. According to Overbeck (p. 55) it existed during the time of Augustus; Coarelli (2002, p. 52) dates the new gate shortly after 80 BC, when Pompeii became a *colonia*.

753. Schultze 287.
754. Stambaugh 188.
755. Säflund 65.
756. Schultze 287.
757. Overbeck 55.
758. Founding these colonies the same rituals were in use, according to Varro, as earlier in Italy: Var. *L.* 5.143 *coloniae nostrae omnes in litteris antiquis scribuntur urbes, quod item conditae ut Roma; et ideo coloniae et urbes conduntur, quod intra pomerium ponuntur.*
759. Schultze 296.
760. Akurgal *passim.*
761. Ach. Tat. 5.1.2 ἐκ τῶν 'Ηλίου πυλῶν εἰς τὰς Σελήνης πύλας. The plural form shows that here there is also talk of a gate with more than one passage.
762. Empereur 53; a reconstruction illustration (pp. 54–5) shows Alexandria with a city wall and single gates.
763. Akurgal 20 and illustrations 30a and b.
764. For example Cologne, cf. Stuart and de Grooth 41; Colchester cf. Wacher 1992, 29.
765. So Cologne was accommodated, besides the three main gates of *cardo* and *decumanus*, with another three gates on the land side (one of them, on the west side, was a double gate) and another three on the river side, so nine in total; Stuart and de Grooth 41; Horn (1987) 465; Klinkenberg 183–97.
766. Schultze 291–3; Donnadieu 94 sqq.
767. Brands 131–2; illustrations 98, 99 (Porta Consolare) and 105–11 (Porta Venere).
768. For a plan of Aosta and approach roads see Walser 62 and Barocelli Tav. IV.
769. Schultze 294 (also the earlier given numbers).
770. Schultze 293–5.
771. Schultze, Taf. XIII, XIV, XVI; figures 3.7 and 3.8; Wacher 1997, 72.
772. Crummy 95–6. He also gives the earlier proposed dates of the defence structures of Colchester.
773. Schultze, Taf. XIV (= figure 3.7); Wacher 1997, 72; Hull 18. There is also the possibility that Balkerne Gate was at first a triumphal arch; van Tilburg 2005, 128–9.
774. Schultze 306.
775. Burginatiumtor is a double gate. It is not exactly known if Veterator was a double gate or three-passage-gate. Maastor was a three-passage-gate; Lehner 1903, 182–4.
776. For a former, no longer valid, plan of the east gates see Böcking 283.
777. Schultze XIII–XVII. *Contra* Cüppers 10, where in illustration 7 an opening is suggested.
778. Heinen 110–11.
779. Dahm 13.
780. Schultze 336. Both towers originally had the same height.
781. Dahm 25.
782. Cüppers 12.
783. Schultze 337; Cüppers 25–6; Dahm 30.
784. Akşit 27.
785. Schultze 330.
786. Schultze 329.
787. Sayar 455–6; Hellenkemper and Hild 96–7 and illustrations 139–40. The former hypothesis that the arch dates from the Empire (maybe a triumphal arch of Marcus Aurelius) has been abandoned.
788. Herzog 846–51.
789. Bible, *Ev. Matt.* 19:24 'And again I say unto you, It is easier for a camel to go through

the eye of a needle, than for a rich man to enter into the kingdom of God'. This was even more the case with a loaded camel. Fawcett 531.

790. Van Es 143; Grénier 1931, 282–4; Drinkwater 151.
791. This was valid for many settlements and cities in Gaul; see note 790 *supra.*
792. Smeesters 15; for the situation of Tongeren see 116–7; 119.
793. Van Es *passim.*
794. Smeesters 34.
795. Willems 81.
796. Cleary 59 and 174. In Rome, Porta Clausa (Aurelian Wall) was bricked up.
797. Porta Nigra was changed into a church; Cüppers 34 sqq.
798. *CIL* VI 1190.
799. Platner and Ashby 411.
800. Platner and Ashby 350; cf. the *Forma Urbis* (Lugli and Gismondi; Lanciani).
801. This gate was later combined with Arch of Drusus, a triumphal arch. So a large gate complex was created; Schultze 343 and Taf. XVII ('Drususbogen'); Platner and Ashby (p. 39) and Lanciani (42) situate the Arch of Drusus more to the north. A combination of city gate and triumphal arch is also found in Pula (Istria); von Hesberg 283–4.
802. Giovagnoli 42 and 101; Platner and Ashby 412. *Contra* Schultze 343. The *Forma Urbis* (Lugli and Gismondi) shows only Porta Appia, Porta Portuensis and Porta Ostiensis as double gates; Lanciani only Porta Portuensis (39).
803. Platner and Ashby 413.
804. Giovagnoli 44–5.
805. Giovagnoli 42–4.
806. Giovagnoli 50.
807. Schultze 342-3; Giovagnoli 105–6.
808. See chapter 3, pp. 101–2 (Trier).
809. Cic. *Inv.* 2.123 *lex aperiri portas noctu vetaret.*
810. Cic. *Inv.* 2.123 *eo, qui* [...] *aperuit quodam in bello et auxilia quaedam in oppidum recepit.*
811. Gross 54–5.
812. Aristid. *Or.* 51.4-5 *νύκτες* [...] *γιγνόμεθα ἐν Κύμῃ* [...] *ἐχώρουν ἔξω πυλῶν.*
813. *Thesaurus Linguae Latinae* 9–10, s.v. 'portarius'.
814. Sil. 2.240 *bone ianitor urbis*; Lyd. *Mag.* 1.38 *ὕπαρχος τὴν πόλιν ἐφύλαττε*, *custos urbis.* Here is talk of politicians, not of gate-keepers.
815. Poll. 1.77 *καλοῦσι* [...] *πυλωροῦντος οἴκημα πυλώριον*; *SEG* 41 86,13 (Rhamnous, 252 BC).
816. There is a very unclear mention of a *πυλωρός* in Aphrodisias (third century), *SEG* 36 970 B 50; *IG* II^2 2292–2310 gives a list of *πυλωροί* in Athens. According to D. Peppa-Delmouzou, the title of *πυλωρός* had declined in the first and second centuries AD; *SEG* 29 153.
817. Mau 236; Etienne 1985, 99; Overbeck 50–1.
818. Matthews 25.
819. Overbeck 53; Mau 238. According to Coarelli (2002, p. 52) there were repairs to this gate after Sulla's attack in 89 BC.
820. Coarelli 2002, 192.
821. See note 752 *supra.*
822. Etienne 1985, 304.
823. Coarelli 2002, 52.
824. Coarelli 2002, 346.
825. Van Binnebeke 142–4.
826. Compare the maps of Stuart and de Grooth (p. 41) and *CCAA Romeins Keulen* (p. 26).

827. Höpken *passim*; Thomas 408.
828. Klinkenberg 179–97; Süßenbach 106–7 (plan backside). For Markt or Mars Tor see Klinkenberg 196; Süßenbach 81.
829. Süßenbach 44-9; the rim of the arch of the central passage is rebuilt in the Römisch-Germanisches Museum, not far away from the original position of the gate; the rim of one of the pedestrian passages is placed *in situ.*
830. Klinkenberg 193-4; Süßenbach 68.
831. Schmidt *passim*; Liesen 238 n. 6; Kühlborn 486, 491.
832. Böcking 218.
833. Böcking 273.
834. According to H.J. Schalles (APX) outside the Burginatiumtor not very much is found, but there have been no important excavations yet.
835. For example Böcking 217, 246, 283; Heimberg and Rieche 7, 34, 60 etc. The black parts of the LVR plan are incorporated in the plan of Kühlborn's article (also showing a three-passage-gate), where the excavated parts are shown in black; Kühlborn 489.
836. Lehner 182–7; Bechert 258–9 with a second reconstruction proposal of a three-passage-gate.
837. *Historia Augusta, Didius Julianus* 1.7 *Inde Belgicam sancte ac diu rexit. Ibi Chaucis, Germaniae populis qui Albim fluvium adcolebant, erumpentibus restitit tumultuariis auxiliis provincialium.*
838. In the fourth century, there were some churches outside the walls; Dahm 12–13.
839. See chapter 3, p. 115 s.v. 'Trier'.
840. Dahm 12, 13, 30 and 31.
841. Wacher 1997, 127; Cleary 46–56.
842. Hull 14–63. For extended research of extramural buildings in Roman Britain see Cleary *passim.*
843. Smeesters 15; Vanvinckenroye 36.
844. Vanvinckenroye 38–40.
845. Vanvinckenroye, attachment.
846. Johnston 51. For more 'bird's foot' examples see Cleary *passim.*
847. Mertens and Vanvinckenroye 57–8.
848. Lugli and Gismondi; Lanciani.
849. See note 83 *supra.*
850. Friedländer 10.
851. Friedländer (p. 23) mentions a figure of 1.5 million inhabitants in the second century, before the plague during the reign of Marcus Aurelius (*Pestis Antoniniana*). Morley (p. 2) quotes 1 million; Wiesehöfer (p. 602) 750,000 to 2 million; Stark 650,000.
852. Morley 2. Stark respectively 400,000 and 150,000.
853. Emperor Claudius is meant.
854. Alf. *dig.* 9.2.52.2 *In clivo capitolino duo plostra onusta mulae ducebant* [. . .] *plostrum a priore percussum retro redierat et puerum cuiusdam obtriverat*; van Binnebeke 123.
855. F. Kolb 388.
856. Mart. 7.61 (p. 133).
857. Suet. *Cl.* 25 *Viatores ne per Italiae oppida nisi aut pedibus aut sella aut lectica transirent, monuit edicto.* See p. 132.
858. Suet. *Nero* 5.1 (pater Neronis) *sed et in viae Appiae vico repente puerum citatis iumentis haud ignarus obtrivit*; Laurence 1999, 154.
859. See note 885 *infra.*
860. See note 420 *supra.* There was also a crowd in Constantinople: Zos. 2.35 ὥστε [. . .] καὶ

μετὰ κινδύνον βαδίζειν διὰ τὴν τῶν ἀνθρώπων καὶ ζῴων πολυπλήθειαν; Pöhlmann 80.

861. Ausonius, *Ordo urbium nobilium* 20.16-17 *latas nomen servare plateas, | tum respondentes directa in compita portas.* Here the forecourts before the gates are meant.
862. B. Wagendorp, *Volkskrant* 28 October 1998, 'Karresporen wijzen uit: Links rijden!'; G. Dyer, *The Telegram* 17 January 1999, 'Is driving on the right right or wrong?' (Internet: http://www.geocities.com/Heartland/Valley/8414/drive-right.htm).
863. Cic. *Fam.* 8.6.5 (Curio) *legemque viariam* [. . .] *iactavit*; App. *BC* 2.27 ὁ δὲ Κουρίων [. . .] εἰσηγεῖτο βαρυτάτας ὁδῶν πολλῶν ἐπισκευάς τε καὶ κατασκευὰς; van Binnebeke 47.
864. Van Binnebeke 47–8.
865. Nap 48–52.
866. Hardy, 'Roman laws and charters' 136; for more discussion and dating see pp. 165–8.
867. Van Binnebeke 44.
868. Homo 420; Marquardt 731; Sonnabend 187; Legras 11, 369–70; Nap *passim*.
869. In summer ± 4.30 a.m. to 17.00 p.m., in winter 7.30 a.m. to 15.00 p.m. (van Binnebeke 121 n. 381).
870. Staccioli speaks of *plostra stercoraria* (p. 20).
871. Liv. 35.9.3 *circa portam Flumentanam etiam collapsa quaedam ruinis sunt.*
872. Var. *R.* 3.2 *extra urbem* [. . .] *eorum aedificia, qui habitant extra portam Flumentanam aut in Aemilianis.*
873. Suet. *Cl.* 18 *Cum Aemiliana pertinacius arderent, in diribitorio duabus noctibus mansit;* Platner and Ashby 151.
874. Scobie 408.
875. Sonnabend 193. In my opinion, it is doubtful if complaints would have had any effect; this is discussed on pp. 167–9.
876. The *Lex Julia Municipalis* only mentions *plostra*, but we must assume that *rhedae* and other travel wagons are also meant, because even triumphators rode in a *plostrum*.
877. For prices of wagons in the *Edictum Diocletiani* see p. 55.
878. Hardy, 'Three Spanish charters and other documents' 35 n. 55.
879. According to the *CIL*, this article also occurs in the *Lex Julia agraria* (59 BC).
880. A *decumanus* is, in fact, an east–west running road, but I presume that here all roads are meant, except the *limites*.
881. A law concerning the foundation of *coloniae*, which is not discussed here.
882. See chapter 1.
883. Van Binnebeke 123 n. 392; Tsujimura 61.
884. Dangerous situations with horses: Lib. *Or.* 1.183 *καὶ* [. . .] δέρματος; 1.216–217 τῆς [. . .] ἵππου.
885. Gesemann (p. 40 n. 26) is of the opinion that it is not clear whether horse and wheeled traffic was forbidden in through-faring streets.
886. Emperor Domitian is meant.
887. Also applicable for the *cursus publicus*: Hudemann 141.
888. Marquardt 731; van Binnebeke 122 n. 388 with commentary. There is no mention that this was applicable at fixed hours; maybe, it was a complete ban without exceptions; Zimmermann 188.
889. Gal. *De cur. rat. per venae sect.* 17 (11.301 K.) *κατά τι χωρίον ἔνθα τῶν ὀχημάτων ἀποβαίνειν εἰσὶν εἰθισμένοι.*
890. Suet. *Cl.* 17 *Currum eius Messalina uxor carpento secuta est.* For other female members of the imperial family see Weber 308.
891. There is no talk of a wagon, but Plautianus is drawn by mules or hinnies: D.C. 77.4.1

καὶ ὃς οὕτως ἠπείχθη [. . .] ὥστε τὰς ἡμιόνους τὰς ἀγούσας αὐτὸν πεσεῖν ἐν τῷ παλατίῳ. For the year 203 see Marquardt 730; he refers *falso* D.C. 76.4.1.

892. Lib. *Or.* 50; Liebeschuetz 51–3.

893. *Historia Augusta, Aurelianus* 5.3–4 *Nam ingrediente* [. . .] *cum in equum transire vellet, quia invidiosum tunc erat vehiculis in civitate uti, equus est ei imperatoris adplicitus.*

894. Weber 309–11 with references.

895. See chapter 1, pp. 37–8.

896. Matthews 25–6.

897. A case of two roads running parallel can be found west of Tivoli; Giuliani 34 (Percorsi A and B). Also in case of a military settlement, an alternative route through a *canaba* or *vicus* could exist; Colenbrander 70–1.

898. *IG* I^3 1 79, 11–14 πλάτος [. . .] hιερά; Zimmermann 184. A wagon with an axle width of about 1.4 m (see also the following note) could, however, pass such a bridge. In the inscription an ἅμαξα is mentioned; this should indicate that such a wagon type was too big to pass the bridge.

899. This is roughly the same as the modern standard gauge of trains, tramways and underground: 1437 mm.

900. Tsujimura 62.

901. The wheels of the wagon in the House of Menander have a diameter of 1.39 m; so about 68 cm to the hub; Tsujimura 61.

902. Tsujimura 62–3; Gesemann 35 and 281.

903. Tsujimura *passim*; the plan on p. 64 is unfortunately not clear concerning the depth of the wheel ruts. Wallace-Hadrill (p. 49) is clearer.

904. Mau 222; Tsujimura 63.

905. Tsujimura 64–9.

906. Mau's opinion is followed by Matthews (p. 24).

907. Street-lighting was not completely unknown in Antiquity. There was street-lighting in Antioch in 450 AD; Forbes 1934, 156; see chapter 2 p. 50–1 on street name plates.

908. Tsujimura 76–7.

909. Tsujimura 64 and 77.

910. Compare the situation of taverns (Wallace-Hadrill 44) and the situation of deep wheel ruts (Wallace-Hadrill 49).

911. See Fig. 4.3.

912. Gesemann 66–7.

913. Gesemann 67–70, where the different obstacles and barricades are described.

914. Wallace-Hadrill 44–9.

915. Zimmermann 192; Gesemann 63. For more information on stepping stones see Gesemann 63–5 and 306 (plan).

916. Tsujimura 85-6.

917. Carmelez 12.

918. See chapter 1, p. 31.

919. In Antiquity: Athens and Milete in the fifth century BC; Nollé 41.

920. Stuart and de Grooth 19; Mertens 46; Rottier 81 and 87.

921. Carmelez 28.

922. Heimberg and Rieche 29; Heimberg 469; see also the map of Xanten, p. xviii.

923. Suet. *Nero* 55 *destinaverat et Romam Neropolim nuncupare.*

924. Plans by Bergmann and Royo situate the Domus Aurea on the Esquiline Hill: Bergmann 20; Royo 239.

925. That the inhabitants condemned the Domus Aurea is further proved by Tacitus: *Ann.* 15.52 *in illa invisa et spoliis civium exstructa domo*; Mart. *Sp.* 2.11–12 *Reddita Roma*

sibi est et sunt te preside, Caesar, | deliciae populi, quae fuerant domini. This last poem is a panegyric on the politics of Vespasian, who gave back the Domus Aurea area to the public, by building the Colosseum. He also states that in Vespasian's time the Domus Aurea was not yet finished according to the original concepts; Platner and Ashby 167.

926. Van der Meer 63–73.
927. Platner and Ashby 168–9.
928. Friedländer 10.
929. Van der Meer 63; Platner and Ashby 170 and 172 with references.
930. Brands and Maischberger 109.
931. *Historia Augusta, Commodus* 15.7 *Urbem incendi iusserat, utpote coloniam suam.*
932. *Historia Augusta, Pertinax* 9.2 *Reformandis viis pecuniam contulit.*
933. For Via Nova see note 266 *supra.*
934. *Historia Augusta, Caracalla* 9.9 *Idem viam novam munivit, quae est sub eius thermis, Antoninianis scilicet, qua pulchrius inter Romanas plateas non facile quicquam invenias*; Platner and Ashby 565. See also the remarks about the width of Via Nova, p. 31, and notes 268 and 269 *supra.*
935. The *Forma Urbis* (Septimius Severus' time) shows many columns along the street side, probably caused by fire-fighting policy; Staccioli 26.
936. Herodianus (Loeb edition) Band 1, 93 n.2.
937. Including Carthage, which had to be renamed to Alexandria Commodiana: *Historia Augusta, Commodus* 17.8 *Ridicule etiam Carthaginem Alexandriam Commodianam togatam appellavit.*
938. Zimmermann 186; Hoepfner and Schwandner illustrations 41 (Rhodes) and 225.
939. Var. *L.* 5.143 *Oppida condebant in Latio Etrusco ritu multi, id est iunctis bobus, tauro et vacca interiore, aratro circumagebant sulcum (hoc faciebant religionis causa die auspicato), ut fossa et muro essent muniti. Terram unde exculpserant, fossam vocabant et introrsum iactam murum. Post ea qui fiebat orbis, urbis principium; qui quod erat post murum, postmoerium dictum, eo usque auspicia urbana finiuntur.* He uses the archaic spelling *postmoerium.* Liv. 1.44.4 *Pomerium, verbi vim solam intuentes, postmoerium interpretantur esse; est autem magis circamoerium, locus quem in condendis urbibus quondam Etrusci, qua murum ducturi erant, certis circa terminis inaugurato consecrabant, ut neque interiore parte aedificia moenibus continuarentur, quae nunc volgo etiam coniungunt, et extrinsecus puri aliquid ab humano cultu pateret soli*; for more discussion and references see von Blumenthal 1868 and Nissen 466–77.
940. Going beyond the *pomerium*, i.e. crossing the wall another way than through the gates, was seen as a war crime and, for example in the case of Romulus and Remus, punishable by the death penalty: Pompon. *dig.* 1.8.11 *Si quis violaverit muros, capite punitur, sicuti si quis transcendet scalis admotis vel alia qualibet ratione. Nam cives romanos alia quam per portas egredi non licet, cum illud hostile et abominandum sit: nam et Romuli frater Remus occisus traditur ob id, quod murum transcendere voluerit.*
941. Nissen 479–80; Ward-Perkins 41.
942. Nissen (p. 480): 'am Fuß der Mauer'.
943. According to Etienne (1985, 298) there was a circular road inside the walls, but I could not find anything on the plans of Coarelli (2002), Tsujimura and Laurence (1994/1996).
944. Nissen 479.
945. Such circular roads, following the *pomerium*, are also found in the British cities Colchester (Wacher 1997, 115), Lincoln (Wacher 1997, 134), Gloucester (Wacher 1997, 153 and 163) and York (Wacher 1997, 166).
946. Varène, figs. 14, 17, 18a and b and 19.
947. Heimberg and Rieche 29; Heimberg 469.

948. Davies 132 and 140; for the position of the *forum* in respect to *cardo* and *decumanus* see Davies 128–9, where the choice of traffic and a fine view is mentioned.
949. In Silchester, after the diversion of the *decumanus*, an obtuse corner was created. In St Albans there was an obtuse corner, caused by following its prehistoric predecessor (Watling Street), comparable with the situation in Xanten; Wacher 1997, 218. In London there were more obtuse angles; the traffic situation at some intersections in London is described in detail by Davies (pp. 138–40). Besides in Trier and Caerwent, also in Tongeren the *cardo* is diverted; in Timgad, St Albans and Cologne the *decumanus.*
950. For example the *castellum* at Valkenburg a/d Rijn: van Es 61; Tac. *Ann.* 1.61 *castra lato ambitu* (Varus). The *principia* is usually blocking the *via praetoria*, causing a bent route; e.g. Colenbrander 2005, 70–1; 74–5.
951. See note 106 *supra.*
952. Liv. 1.44.3 (Servius Tullius) *Addit duos colles, Quirinalem Viminalemque; inde deinceps auget Esquilias; ibique ipse, ut loco dignitas fieret, habitat. Aggere et fossis et muro circumdat urbem; ita pomerium profert*; Tac. *Ann.* 12.24 *forumque et Capitolium non a Romulo, sed a Tito Tatio additum urbi credidere. Mox pro fortuna pomerium auctum*; von Blumenthal 1872–3.
953. Sen. *De brevitate vitae* 13.8 *Sullam ultimum Romanorum protulisse pomerium, quod numquam provinciali sed Italico agro adquisito proferre moris apud antiquos fuit.*
954. Tac. *Ann.* 12.23 *pomerium urbis auxit Caesar, more prisco, quo iis qui protulere imperium etiam terminos urbis propagare datur.*
955. Lugli and Gismondi.
956. Hor. *S.* 1.8.15 *Aggere in aprico spatiari.* Here are the Gardens of Maecenas; Platner and Ashby 269.
957. Claudius: Tac. *Ann.* 12.24 *Et quos tum Claudius terminos posuerit, facile cognitu et publicis actis perscriptum*; Gel. 13.14.7 *extra pomerium exclusum, post auctore divo Claudio receptum et intra pomerii fines observatum*; Vespasian: *CIL* VI 31538a–c; Hadrian: *CIL* VI 31539a–c.
958. *Historia Augusta, Aurelianus* 21.9–11 *adhibito consilio senatus muros urbis Romae dilatavit. Nec tamen pomerio addidit eo tempore sed postea. Pomerio autem neminem principum licet addere nisi eum qui agri barbarici aliqua parte Romanam rem publicam locupletaverit. Addidit autem Augustus, addidit Traianus, addidit Nero.* For the expansion by Augustus, Trajan and Nero see von Blumenthal 1873–5. For an overview of the steady expansion from Romulus to Aurelian see Simonelli 151–60.
959. Palmer 1980, 217–18 and 226.
960. See note 957 *supra* (Claudius).
961. *CIL* X 3825.
962. Var. *L.* 5.143 *Cippi pomeri stant et circum Ariciam et circum Romam.*
963. In a sense, every city wall can be seen as a circular road. Pedestrians can make a circular tour around the city.
964. See chapter 3, pp. 87–8.
965. De Wit *passim.* Exceptions were Leiden, Gouda and Utrecht amongst others. For British cities *c.* 1800 see Lobel *passim.*
966. Sometimes a trader could get a promotion to *decurio*; in Nijmegen there was a *decurio* Hilarus. Probably there was no choice; Willems 43–4.
967. Von Stephan 17.
968. Laurence 1999, 154 (Minturnae).

BIBLIOGRAPHY

1. Ancient sources

Greek sources are abbreviated according to the list of abbreviations of H.G. Liddell and R. Scott (eds), *A Greek–English Lexicon*, Oxford 1978[9].

Latin source are abbreviated according to the list of abbreviations of P.G.W. Glare (ed.), *Oxford Latin Dictionary*, Oxford 1982.

Other ancient sources are not abbreviated.

L. = Loeb edition (Cambridge, MA)
T. = Teubner edition (Leipzig)
OCT = Oxford Classical Texts (Oxford)

Achilles Tatius, [*Leucippe et Clitophon*], ed. Gaselee **L.**
(Aelius) Aristides, *Orationes XVII–LIII*, ed. Keil, 1898 [Vol. II].
Agathias, *Historiae*, ed. Keydell, 1967.
Alfenus Varus, *Digestae*, in *Corpus Iuris Civilis.*
Ambrosius, *Expositio psalmi CXVIII*, ed. Schenkl, 1913 [CSEL 62].
Ammianus Marcellinus, *Res gestae*, ed. Rolfe **L.**
Anonymus, *De viris illustribus*, ed. Sherwin Jr, 1973.
Appianus, *Bella Civilia*, ed. Evelyn White **L.**
Apuleius, *Florida*, ed. Helm **T.**
——, *Metamorphoses*, ed. Adlington **L.**
Arcadius, *Digestae*, in *Corpus Iuris Civilis.*
Aristophanes, *Vespae*, ed. Rogers **L.**
Aristoteles, *Atheniensium respublica*, ed. Rackham **L.**
Arrianus, *Acies contra Alanos*, ed. Roos **T.** [Vol. II].
——, *Anabasis*, ed. Robson **L.**
Athenaeus, *Dipnosophistae*, ed. Kaibel **T.**
Augustus, *Monumentum Ancyranum*, ed. Shipley **L.**
Ausonius, *Epistulae*, ed. Evelyn White **L.**
——, *Ordo urbium nobilium*, ed. Evelyn White **L.**
Bible, ed. R. Carroll and S. Prickett, Oxford etc. 1997 [Authorized King James Version with Apocrypha].
Caesar, *De bello civile*, ed. Peskett **L.**
Cassiodorus, *Variae*, ed. Fridh, 1973 [CCSL 96].

Cato, *De agricultura*, ed. Hooper and Ash **L**.
Catullus, *Carmina*, ed. Cornish **L**.
Celsus, *De medicina*, ed. Spencer **L**.
Cicero, *De inventione*, ed. Hubbell **L**.
——, *De natura deorum*, ed. Rackham **L**.
——, *De officiis*, ed. Miller **L**.
——, *De provinciis consularibus*, ed. Gardner **L**.
——, *Epistulae ad Atticum*, ed. Winstedt **L**.
——, *Epistulae ad familiares*, ed. Williams **L**.
——, *Epistulae ad Quintum fratrem*, ed. Cary **L**.
——, *In Pisonem*, ed. Watts **L**.
——, *Philippicae*, ed. Ker **L**.
——, *Pro Fonteio*, ed. Watts **L**.
——, *Pro lege Manilia*, ed. Hodge **L**.
——, *Pro Milone*, ed. Watts **L**.
——, *Pro Sestio*, ed. Gardner **L**.
——, *Tusculanae disputationes*, ed. King **L**.
Claudianus, *De bello Gothico*, ed. Platnauer **L**.
——, *Opera minora*, ed. Platnauer **L**.
Codex Iustinianus, in *Corpus Iuris Civilis*.
Codex Theodosianus, ed. Mommsen and Meyer, 1905.
Columella, *De re rustica*, ed. Ash **L**.
Corpus Iuris Civilis, ed. Mommsen and Krüger, 1928.
Dio Cassius, *Historiae Romanae*, ed. Cary **L**.
Diocletianus, *Edictum Diocletiani de pretiis*, ed. Lauffer, 1971.
Diodorus Siculus, *Bibliotheca historica*, ed. Geer **L**.
Dionysius Halicarnassensis, *Antiquitates Romanae*, ed. Cary **L**.
Epictetus, *Diatribae*, ed. Schenkl **T**.
Euripides, *Electra*, ed. Way **L**.
Eutropius, *Breviarium ab urbe condita*, ed. Santini **T**.
Festus, *De verborum significatu*, ed. Lindsay **T**.
Frontinus, *De Aquis Vrbis Romae*, ed. Bennett **L**.
Gaius, *Digestae*, in *Corpus Iuris Civilis*.
Galenus, *De curandi ratione per venae sectione*, ed. Kühn.
——, *De praecognitione ad Epigenum*, ed. Kühn.
Gellius, *Noctes Atticae*, ed. Rolfe **L**.
Herodianus, ed. Whittaker **L**.
Herodotus, *Historiae*, ed. Godley **L**.
Hieronymus, *Epistulae*, ed. Hilberg, 1996 [CSEL 54–6].
Hippocrates, *De aëre aquis locis*, ed. Jones **L**.
Historia Augusta, ed. Magie **L**.
Horatius, *Epistulae*, ed. Fairclough **L**.
——, *Sermones*, ed. Fairclough **L**.
Hyginus, *Constitutio limitum*, in *Corpus Agrimensorum Romanorum*, ed. Thulin **T**.
——, *De limitibus*, in *Corpus Agrimensorum Romanorum*, ed. Thulin **T**.
——, *De metatione castrorum*, ed. Grillone **T**.
Javolenus, *Digestae*, in *Corpus Iuris Civilis*.
Isidorus, *Origines* sive *Etymologiae*, ed. Lindsay, 1911.

Itinerarium Burdigalense, ed. Geyer and Cuntz, 1965 [CCSL 175–6].
Josephus, *Bellum Judaicum*, ed. Thackeray **L**.
Julianus, *Misopogon*, ed. Wright **L**.
Juvenalis, *Saturae*, ed. Ramsay **L**.
Leges XII Tabularum, ed. Düll, 1953.
Lex Julia Municipalis = *CIL* I^2 593.
Libanius, *Orationes*, ed. Förster **T**.
Liber Coloniarum, ed. Lachmann, 1848.
Livius, *Ab urbe condita*, ed. Forster, Schlesinger, Moore and Sage **L**.
——, *Periochae*, ed. Rossbach **T**.
Lydus, *De magistratibus populi Romani*, ed. Wünsch **T**.
Macrobius, *Saturnalia*, ed. Willis **T**.
Martialis, *De spectaculis*, ed. Shackleton Bailey **L**.
——, *Epigrammata*, ed. Shackleton Bailey **L**.
Nonius Marcellus, *De conpendiosa doctrina*, ed. Lindsay **T**.
Notitia dignitatum, ed. Seeck, 1962.
Optatus Milevitanus, *Libri VII, appendix*, ed. Ziwsa, 1893 [CSEL 26].
Ovidius, *Fasti*, ed. Frazer **L**.
Panegyrici Latini, ed. Baehrens **T**.
Papinianus, *Digestae*, in *Corpus Iuris Civilis*.
Paulus, *Digestae*, in *Corpus Iuris Civilis*.
——, *Epitome*, in Festus, *De verborum significatu*, ed. Lindsay **T**.
——, *Sententiae*, ed. Huschke **T**. [Iurisprundentia anteiustiniana II.1].
Pausanias, *Graeciae descriptio*, ed. Jones **L**.
Petronius, *Satyricon*, ed. Heseltine **L**.
Philo, *Legatio ad Gaium*, ed. Colson **L**.
Plato, *Hipparchus*, ed. Burnet **OCT**.
Plautus, *Amphitruo*, ed. Nixon **L**.
——, *Pseudolus*, ed. Nixon **L**.
Plinius Maior, *Naturalis Historia*, ed. Detlefsen, 1866–82.
——, *Naturalis Historia*, ed. Rackham **L**.
Plinius Minor, *Epistulae*, ed. Melmoth **L**.
——, *Epistulae ad Traianum*, ed. Malmoth **L**.
Plutarchus, *C. Gracchus*, ed. Perrin **L**.
——, *Caesar*, ed. Perrin **L**.
——, *Moralia*, ed. Fowler and Helmbold **L**.
——, *Pompeius*, ed. Perrin **L**.
Pollux, *Onomasticon*, ed. Bethe **T**.
Polybius, *Historiae*, ed. Paton **L**.
Pomponius, *Digestae*, in *Corpus Iuris Civilis*.
Procopius, *De bello Gothico*, ed. Dewing **L**.
——, *Historia Arcana* sive *Anecdota*, ed. Dewing **L**.
Quintilianus, *Declamationes (minores)*, ed. Winterbottom, 1984.
Rutilius Namatianus, *De reditu suo*, ed. Duff and Duff **L**.
Scaevola, *Digestae*, in *Corpus Iuris Civilis*.
Seneca Minor, *Apocolocyntosis*, ed. Rouse **L**.
——, *De brevitate vitae*, ed. Basore **L**.
——, *De clementia*, ed. Basore **L**.

——, *De ira*, ed. Basore **L.**
——, *Epistulae*, ed. Gummere **L.**
Servius, *In Vergilium Commentarius*, ed. Thilo and Hagen **T.**
Silius Italicus, *Punica*, ed. Duff **L.**
Statius, *Silvae*, ed. Mozley **L.**
Strabo, *Geographia*, ed. Jones **L.**
Suetonius, *Augustus*, ed. Rolfe **L.**
——, *Caligula*, ed. Rolfe **L.**
——, *Claudius*, ed. Rolfe **L.**
——, *Iulius*, ed. Rolfe **L.**
——, *Nero*, ed. Rolfe **L.**
——, *Tiberius*, ed. Rolfe **L.**
——, *Vespasianus*, ed. Rolfe **L.**
Symmachus, *Epistulae*, ed. Callu, 1972–2002 [Budé].
Tacitus, *Annales*, ed. Jackson **L.**
Terentius, *Adelphi*, ed. Sargeaunt **L.**
——, *Eunuchus*, ed. Sargeaunt **L.**
Tertullianus, *De anima*, ed. Waszink, 1947.
Tibullus, *Elegiae*, ed. Postgate **L.**
Traianus, in Plinius Minor, *Epistulae.*
Ulpianus, *Digestae*, in *Corpus Iuris Civilis.*
Varro, *De lingua Latina*, ed. Kent **L.**
——, *De re rustica*, ed. Hooper and Ash **L.**
Vegetius, *Epitoma rei militaris*, ed. Müller, 1997.
Vergilius, *Georgica*, ed. Fairclough **L.**
Vitruvius, *De architectura*, ed. Granger **L.**
Xenophon, *Anabasis*, ed. Brownson **L.**
Zosimus, *Historia nova*, ed. Mendelssohn **T.**

2. Epigraphical sources

AE — *l'Année Epigraphique: Revue des publications épigraphiques relatives l'antiquité romaine.*
CIG — *Corpus Inscriptionum Graecarum.*
CIL — *Corpus Inscriptionum Latinarum.*
IG — *Inscriptiones Graecae.*
ILLRP — *Inscriptiones Latinae Liberae Rei Publicae.*
ILS — *Inscriptiones Latinae Selectae.*
IvEph. — *Die Inschriften von Ephesos.*
P. Oxy. — *Papyri Oxyrhinchi.*
SEG — *Supplementum Epigraphicum Graecum.*
Supp. Ital. — *Supplementa italica-imagines: supplementi fotografici ai volumi italiani del CIL.*

3. Modern authors

Adams, C. and R. Laurence, *Travel and Geography in the Roman Empire.* London and New York 2001.
Akşit, I., *Pamukkale Hierapolis Aphrodisias.* Istanbul 1993 [Dutch version].

Akurgal, E., *Ancient Civilizations and Ruins of Turkey: From Prehistoric Times until the End of the Roman Empire.* Istanbul 1973[3] [transl. J. Whybrow and M. Emre].

André, J.-M. and M.-F. Baslez, *Voyager dans l'Antiquité.* Paris 1993.

Barocelli, P., *Forma Italiae: Regio XI Transpadana. Vol. I: Avgvsta Praetoria.* Rome 1948.

Basso, P. *et al. Via per montes excisa: strade in galleria e passaggi sotterranei nell'Italia romana.* Rome 1997.

Baume, P. la, *Colonia Agrippinensis: Kurzer Rundgang durch das römische Köln.* Cologne 1958.

Bechert, T., 'Römische Lagertore und ihre Bauinschriften', *Bonner Jahrbücher* 171 (1971) 201–87.

Bender, H., *Römischer Reiseverkehr: Cursus publicus und Privatreisen.* Stuttgart 1978.

——, 'Transportwege, Mittel des Transports und Nachrichtenwesen in der römischen Antike', *Humanistische Bildung* 6 (1983) 137–77 [with the title 'Der Mensch in seiner Umwelt'].

——, 'Verkehrs- und Transportwesen in der römischen Kaiserzeit', in: H. Jankuhn, W. Kimmig and E. Ebel (eds), *Untersuchungen zu Handel und Verkehr der vor- und frühgeschichtlichen Zeit in Mittel- und Nordeuropa: Bericht über die Kolloquien der Kommission für die Altertumskunde Mittel- und Nordeuropas in den Jahren 1980 bis 1983*, V, Göttingen 1989, 108–54.

Bergmann, M., *Der Koloß Neros, die Domus Aurea und der Mentalitätswandel im Rom der frühen Kaiserzeit.* Mainz 1994 [Trierer Winckelmannsprogramme Heft 13].

Binnebeke, M.C. van, *Juridische aspecten van Romeinse stedelijke infrastructuur.* Utrecht 1997.

Bloemers, J.H.F., 'Twenty-five years ROB research in Roman Limburg', *Berichten van de Rijksdienst voor het Oudheidkundig Bodemonderzoek* 23 (1973) 237–58.

Blumenthal, V., 'Pomerium', in: A. Pauly *et al.* (eds), *Paulys Realencyclopädie der classischen Altertumswissenschaft.* Stuttgart and Munich 1903–78, 1867–76 [Vol. 41[2]].

Böcking, W., *Der Niederrhein zur römischen Zeit: Archäologische Ausgrabungen in Xanten.* Kleve 1987.

Boëthius, A. and J.B. Ward-Perkins, *Etruscan and Roman Architecture.* Harmondsworth 1970.

Bonomi Ponzi, L., 'Nuove conoscenze sulle infrastrutture della via Flaminia in Umbria', in: L. Quilici and S. Quilici Gigli (eds), *Strade Romane: percorsi e infrastrutture.* Rome 1997, 155–66 [Atlante tematico di topografia antica 2 – 1993].

Borger, H., 'Die Ausgrabungen in der Dom-Immunität zu Xanten in den Jahren 1957 und 1958 (Vorbericht 1)', *Bonner Jahrbücher* 160 (1960) 313–41.

Brands, G., *Republikanische Stadttore in Italien.* Oxford 1988 [BAR British Series 0143-3067; 458].

Brands, G. and M. Maischberger, 'Der Tempel des Hercules Invictus, die Porta Trigemina und die Porta Triumphalis', *Rivista di Archeologia* 19 (1995) 102–20.

Brodersen, K., 'The presentation of geographical knowledge for travel and transport in the Roman world: *itineraria non tantum adnotata sed etiam picta*', in: C. Adams and R. Laurence (eds), *Travel and Geography in the Roman Empire*, London and New York 2001, 7–21.

Bulle, H., 'Geleisestraßen des Altertums', *Sitzungsberichte der Bayerischen Akademie der Wissenschaften: Philosophisch-historische Klasse* (1947) 2: 1–133 + ill.

Buren, A.W. van, 'Pons', in: A. Pauly *et al.* (eds), *Paulys Realencyclopädie der classischen Altertumswissenschaft.* Stuttgart and Munich 1903–78, 2428–37 [Suppl. vol. 21[2]].

Cagnat, R., 'Portorium, portus', in: C. Daremberg and E. Saglio (eds), *Dictionnaire des antiquités grecques et romaines d'après les textes et les monuments.* Paris 1877–1919, 586–94 [Vol. IV.1].

Calzolari, M., 'Introduzione allo studio della rete stradale dell'Italia romana: l'*Itinerarium Antonini*', *Atti della Accademia Nazionale dei Lincei Memorie* 7 (1996) 4: 369–520 [Serie 9].

Cambridge Ancient History, Cambridge 2000 [Vol. XI].

Carmelez, J.C., *Bavay, Romeinse stad.* Tongeren 1983 [transl. M.M. Moors].

Cartellieri, W., 'Die römischen Alpenstrassen über den Brenner, Reschen-Scheideck und Plöckenpass mit ihren Nebenlinien', *Philologus* Supp. 18 (1926) 1: 1–186 + ill.

Casson, L., *Travel in the Ancient World.* London 1974.

Castritius, H., 'Zum höfischen Protokoll in der Tetrarchie', *Chiron* 1 (1971) 365–76.

Catalogue of the Greek Papyri in the John Rylands Library, Manchester, Manchester 1911– [Vol. IV].

CCAA: Romeins Keulen, Leiden [unpublished reader Universiteit Leiden].

Chapot, V., 'Via', in: C. Daremberg and E. Saglio (eds), *Dictionnaire des antiquités grecques et romaines d'après les textes et les monuments.* Paris 1877–1919, 777–820 [Vol. V].

Chevallier, R., *Les voies romaines.* Paris 1997.

—— *Voyages et déplacements dans l'empire romain.* Paris 1988.

Claessen, H.J.M., 'Zover de weg reikt', in: H.J.M. Claessen (ed.), *Over de weg: de betekenis van infrastructurele voorzieningen in en voor vroege staten.* Leiden 1995, 149–70.

Clauss, M., 'Heerwesen', in: *Realencyklopädie für Antike und Christentum.* Stuttgart 1950–, 1073–1113 [Vol. XIII].

Clavel-Lévêque, M., 'Centuriation, géométrie et harmonie: le cas du Biterrois', in: J.-Y. Guillaumin (ed.), *Mathématiques dans l'Antiquité.* Saint-Etienne 1992, 161–84.

Cleary, S.E., *Extra-mural Areas of Romano-British Towns.* Oxford 1987 [BAR British Series 0143-3032; 169].

Coarelli, F., 'La consistenza della città nel periodo imperiale', in: *La Rome impériale, démographie et logistique: Actes de la table ronde (Rome, 25 mars 1994).* Rome 1997, 89–109.

——, (ed.), *Pompeii.* New York 2002.

Colenbrander, B. (ed.), *Limes Atlas.* Rotterdam 2002.

——, *et al.* (eds), *De gewenste geschiedenis: pleidooi voor het Limes-programma.* Utrecht 2003.

Collinet, P., 'Vehiculum', in: C. Daremberg and E. Saglio (eds), *Dictionnaire des antiquités grecques et romaines d'après les textes et les monuments.* Paris 1877–1919, 668–9 [Vol. V].

Cozzo, G., *Ingegneria romana: maestranza romane, strutture preromane, strutture romane.* Rome 1928.

Crouwel, J.H., 'Vervoermiddelen over land', *Hermeneus* 73 (2001) 2: 115–21.

Crummy, P., 'Colchester: making towns out of fortresses and the first urban fortifications in Britain', in: H. Hurst (ed.), *The 'Coloniae' of Roman Britain: New Studies and a Review.* Portsmouth, RI 1999, 88–100 [papers of the conference held at Gloucester on 5–6 July 1997].

Cüppers, H., *Trier: Porta Nigra.* Mainz 1980 [Führer der Verwaltung der staatlichen Schlösser Rheinland-Pfalz 10].

Czysz, W., '350 Meilen vom Po zur Donau – die römische Staatsstraße *Via Claudia Augusta*', in: H. Koschik (ed.), '*Alle Wege führen nach Rom . . .*': *Internationales Römerstraßenkolloquium Bonn.* Pulheim Brauweiler 2004, 101–16 [Materialien zur Bodendenkmalpflege im Rheinland 16].

Dahm, L. *Trier: Stadt und Leben in römischer Zeit.* Trier 1991.

Davies, H., *Roads in Roman Britain.* Stroud 2002.

DeLaine, J., *The Baths of Caracalla: A Study in the Design, Construction, and Economics of Large-scale Building Projects in Imperial Rome.* Portsmouth, RI 1997 [Journal of Roman Archaeology Supplementary Series, 1063-4304; no. 25].

Dilke, O.A.W., *The Roman Land Surveyors: An Introduction to the Agrimensores.* Newton Abbot 1971.

Donnadieu, A., *La Pompéi de la Provence: Fréjus: Forvm Ivlii.* Paris 1928^2.

Drinkwater, J.F., *Roman Gaul: The Three Provinces, 58 BC–AD 260*. London 1983.

Drijvers, J.W., 'De pelgrimage van Egeria', *Hermeneus* 73 (2001) 2: 174–9.

Eck, W., '*Cura viarum* und *cura operum publicorum* als kollegiale Ämter im frühen Prinzipat', *Klio* 74 (1992) 237–45.

——, *Die staatliche Organisation Italiens in der hohen Kaiserzeit*. Munich 1979 [Vestigia: Beiträge zur alten Geschichte 28].

Eitrem, S. 'Hermai', in: A. Pauly *et al.* (eds), *Paulys Realencyclopädie der classischen Altertumswissenschaft*. Stuttgart and Munich 1903–78, 696–708 [Vol. VIII].

Empereur, J.-Y., *Alexandria Rediscovered*. Paris 1998 [transl. M. Maehler].

Erdkamp, P., 'Legers in beweging: troepenverplaatsingen over land en zee in het Romeinse keizerrijk', *Hermeneus* 73 (2001) 2: 136–45.

Es, W.A. van, *De Romeinen in Nederland*. Haarlem 1981[2].

Esch, A., *Römische Straßen in ihrer Landschaft*. Mainz 1997.

Etienne, R., '*Extra Portam Trigeminam*: espace politique et espace économique à l'*Emporium* de Rome', in: Centre National de la Recherche Scientifique Paris and l'Ecole Française de Rome (eds), *L'urbs: espace urbain et histoire (Ier siècle av. J.-C. – IIIe siècle ap. J.-C. Actes du colloque international organisé par le Centre National de la Recherche Scientifique et l'Ecole Française de Rome (Rome, 8–12 mai 1985)*, Rome 1987, 235–49 [Collection de l'Ecole Française de Rome 98].

——, *Het dagelijks leven in Pompeji*. Baarn 1985 [Dutch translation of *La vie quotidienne à Pompéi*, Paris 1966].

Fawcett, S.V., 'Needle', in: G.A. Buttrick *et al.* (eds), *The Interpreter's Dictionary of the Bible: An Illustrated Encyclopedia Identifying and Explaining All Proper Names and Significant Terms and Subjects in the Holy Scriptures, including the Apocrypha, with Attention to Archaeological Discoveries and Researches into the Life and Faith of Ancient Times*. New York 1962–76, 531 [Vol. III].

Finley, M.I., *The Ancient Economy*. London 1985.

Forbes, R.J., *Man the Maker: A History of Technology and Engineering*. New York 1950 [Dutch translation: *De mens bouwt zich een wereld: Vijfduizend jaar techniek*. Amsterdam 1952].

——, *Notes on the History of Ancient Roads and Their Construction*. Amsterdam 1934 [Allard Pierson Stichting: Archaeologisch-historische bijdragen].

Fougères, G., 'Porta', in: C. Daremberg and E. Saglio (eds), *Dictionnaire des antiquités grecques et romaines d'après les textes et les monuments*. Paris 1877–1919, 581–4 [Vol. IV.2].

Frei-Stolba, R., 'Strassenunterhalt und Strassenreinigung in Rom: zu einigen Sectionen der Tabula Heracleensis', *Historia* 60 (1989) 25–37 [with the title 'Labor omnibus unus'].

Friedländer, L., *Sittengeschichte Roms*. Vienna 1934.

Garnsey, P., *Famine and Food Supply in the Graeco-Roman World: Responses to Risk and Crisis*. Cambridge 1988.

——, 'Famine in Rome', in: P. Garnsey and C.R. Whittaker (eds), *Trade and Famine in Classical Antiquity*. Cambridge 1983, 56–65 [Cambridge Philological Society Suppl. 8].

Gesemann, B., *Die Straßen der antiken Stadt Pompeji: Entwicklung und Gestaltung*. Frankfurt am Main 1996 [Europäische Hochschulschriften, Reihe XXXVIII Archäologie 56; PhD thesis Marburg 1995].

Giebel, M., *Reisen in der Antike*. Düsseldorf 1999.

Giovagnoli, A., *Le porte di Roma*. Rome 1973.

Giuliani, C.F., 'Il clivus tiburtinus ed il sistema stradale connesso', in: L. Quilici and S. Quilici Gigli (eds), *Tecnica Stradale Romana.* Rome 1999, 33–8 [Atlante tematico di topografia antica 1].

Grabherr, G., 'Methodische Ansätze der Römerstraßenforschung im Alpenraum am Beispiel der *Via Claudia Augusta*, in: H. Koschik (ed.), *'Alle Wege führen nach Rom . . .': Internationales Römerstraßenkolloquium Bonn.* Pulheim Brauweiler 2004, 117–130 [Materialien zur Bodendenkmalpflege im Rheinland 16].

Grénier, A., *Manuel d'archéologie gallo-Romaine.* Paris 1931 [Pt 1: Géneralités; Travaux militaires].

——, *Manuel d'archéologie gallo-Romaine.* Paris 1934 [Pt 2: L'archéologie du sol, 1: Les routes].

Grewe, K., *Licht am Ende des Tunnels: Planung und Trassierung im antiken Tunnelbau.* Mainz 1998.

Gross, W.H., 'Bildungsreisen in der römischen Kaiserzeit', *Humanistische Bildung* 6 (1983) 47–84. [with the title 'Der Mensch in seiner Umwelt']

Grünewald, T., *Räuber, Rebellen, Rivalen, Rächer: Studien zu* latrones *im römischen Reich.* Stuttgart 1999 [Forschungen zur antiken Sklaverei 31].

Hagen, J., *Erläuterungen zum Geschichtlichen Atlas der Rheinprovinz. Band 8: Römerstrassen der Rheinprovinz.* Bonn 1931^2 [Publikationen der Gesellschaft für Rheinische Geschichtskunde 12].

Hagesteijn, R., 'Communicatie en infrastructuur in Thaise vroege staten', in: H.J.M. Claessen (ed.), *Over de weg: de betekenis van infrastructurele voorzieningen in en voor vroege staten.* Leiden 1995, 93–106.

Halfmann, H., *Itinera principum: Geschichte und Typologie der Kaiserreisen im römischen Reich.* Stuttgart 1986.

Hardy, E.G., 'Roman laws and charters', in: E.G. Hardy (ed. and trans.), *Roman Laws and Charters and Three Spanish Charters and Other Documents.* New York 1975 [repr. 1912].

——, 'Three Spanish charters', in: E.G. Hardy (ed. and trans.), *Roman Laws and Charters and Three Spanish Charters and Other Documents.* New York 1975 [repr. 1912].

Harsh, P.W., 'Angiportum, Platea, and Vicus', *Classical Philology* 32 (1937) 1: 44–58.

Hascher, M. and H. Sonnabend, 'Konkurrierende Verkehrssysteme in der Antike?', in: E. Olshausen and H. Sonnabend, *Stuttgarter Kolloquium zur historischen Geographie des Altertums 7, 1999: Zu Wasser und zu Land: Verkehrswege in der antiken Welt.* Stuttgart 2002, 17–28 [Geographica Historica 17].

Heck, A. van, *Breviarivm Vrbis Romae Antiqvae.* Lugdunum Batavorum and Rome 1977.

Heichelheim, F.M., 'Bevölkerungswesen', in: A.F. von Pauly, K. Ziegler and W. Sontheimer (eds), *Der kleine Pauly: Lexikon der Antike.* Stuttgart 1964–75, 879 [Vol. II].

Heimberg, U., 'Colonia Ulpia Traiana: die früheste Keramik aus der Forumausgrabung', *Bonner Jahrbücher* 187 (1987) 411–74.

Heimberg, U. and A. Rieche, *Colonia Ulpia Traiana: die römische Stadt: Planung Architektur Ausgrabung.* Cologne 1998.

Heinen, H., *Trier und das Trevererland in römischer Zeit.* Trier 1985 [2000 Jahre Trier, Band 1].

Hellenkemper, H. and F. Hild, *Neue Forschungen in Kilikien*, Vienna 1986 [Denkschriften Philosophisch-historische Klasse Band 186; Veröffentlichungen der Kommission für die Tabula Imperii Byzantini Band 4].

Heres, T.L., 'Rustpauze onderweg: herbergen en wisselstations', *Hermeneus* 73 (2001) 2: 87–101.

Herzig, H.E., 'Die antiken Verkehrswege der Schweiz: neuere Forschungen zu den römischen Straßen', in: E. Olshausen and H. Sonnabend, *Stuttgarter Kolloquium zur Historischen Geographie des Altertums 7, 1999: Zu Wasser und zu Land: Verkehrswege in der antiken Welt.* Stuttgart 2002, 9–16 [Geographica Historica 17].

——, 'Probleme des römischen Straßenwesens: Untersuchungen zur Geschichte und Recht', *Aufstieg und Niedergang der Römischen Welt* 2.1 (1974) 593–648.

Herzog, Z., 'Fortifications', in: D.N. Freedman (ed.), *The Anchor Bible Dictionary*, New York 1992, 844–52 [Vol. II].

Hesberg, H. von, 'Bogenmonumenten der frühen Kaiserzeit und des 2. Jahrhunderts n. Chr. Von Ehrenbogen zum Festtor', in: H.-J. Schalles, H. von Hesberg and P. Zanker (eds), *Die römische Stadt im 2. Jahrhundert n. Chr.: der Funktionswandel des öffentlichen Raumes*, Cologne 1992, 277–99 [Xantener Berichte 2].

Hindle, B.P., *Medieval Roads.* Aylesbury 1982.

Hindley, G., *A History of Roads.* London 1971.

Hinrichs, F.T., 'Der römische Straßenbau zur Zeit der Gracchen', *Historia* 16 (1967) 162–76.

Hoepfner, W. and E.-L. Schwandner, *Haus und Stadt im klassischen Griechenland.* Munich 1994 [Wohnen in der klassischen Polis Band 1].

Holland, L.A., *Janus and the Bridge.* Rome 1961.

Homo, L., *Rome impériale et l'urbanisme dans l'antiquité*, Paris 1971.

Höpken, C., 'Die Produktion römischer Gefässkeramik in Köln', *Kölner Jahrbuch* 32 (1999) 759–70.

Horn, H.G. *et al.* (eds), *Die Römer in Nordrhein-Westfalen.* Stuttgart 1987.

——, (eds), *Fundort Nordrhein-Westfalen: Millionen Jahre Geschichte.* Cologne 2000.

Hudemann, E.E., *Geschichte des römischen Postwesens während der Kaiserzeit.* Berlin 1875.

Hull, M.R. *et al. Roman Colchester.* Oxford 1958.

Humm, M., 'Appius Claudius Caecus et la construction de la Via Appia', *Mélanges de l'Ecole Française de Rome Antiquité* 108 (1996) 2: 693–746.

Hünemörder, C., 'Giraffe', in: H. Cancik and H. Schneider (eds), *Der neue Pauly: Enzyklopaedie der Antike.* Stuttgart 1996–, 1075 [Vol. IV].

——, 'Nashorn', in: H. Cancik and H. Schneider (eds), *Der neue Pauly: Enzyklopaedie der Antike.* Stuttgart 1996–, 721–2 [Vol. VIII].

——, 'Nilpferd', in: H. Cancik and H. Schneider (eds), *Der neue Pauly: Enzyklopaedie der Antike.* Stuttgart 1996–, 946–7 [Vol. VIII].

Johnston, D.E., *An Illustrated History of Roman Roads in Britain.* Bourne End 1979.

Junkelmann, M., *Panis militaris: die Ernährung des römischen Soldaten oder der Grundstoff der Macht.* Mainz 1997².

Klaffenbach, G., *Die Astynomeninschrift von Pergamon.* Berlin 1954 [Abhandlungen der Deutschen Akademie der Wissenschaften zu Berlin 1953, 6].

Kleberg, T., *In den Wirtshäusern und Weinstuben des antiken Rom.* Berlin 1963 [transl. W. Braun].

Klinkenberg, J., 'Das römische Köln', in: P. Clemen (ed.), *Die Kunstdenkmäler der Stadt Köln*, II. Abt., Düsseldorf 1906 [Die Kunstdenkmäler der Rheinprovinz 6 II].

Knauss, J., 'Furt oder Brücke: hydrotechnische Aspekte des mykenischen Straßenbaus in der Argolis', in: E. Olshausen and H. Sonnabend, *Stuttgarter Kolloquium zur historischen Geographie des Altertums 7, 1999: Zu Wasser und zu Land: Verkehrswege in der antiken Welt.* Stuttgart 2002, 323–59 [Geographica Historica 17].

Koepp, F., *Die Römer in Deutschland.* Bielefeld 1912².

Kolb, A., 'Transport and communication in the Roman state: The cursus publicus', in: C. Adams and R. Laurence (eds), *Travel and Geography in the Roman Empire*, London and New York 2001, 95–105.

——, *Transport und Nachrichtentransfer im römischen Reich*, Berlin 2000 [*Klio* Neue Folge Band 2].

Kolb, F., *Rom: Die Geschichte der Stadt in der Antike*. Munich 1995.

Korres, M., *From Pentelicon to the Parthenon: The Ancient Quarries and the Story of a Half-worked Column Capital of the First Marble Parthenon*. Athens 1995.

Kudlien, F., *Der griechische Arzt im Zeitalter des Hellenismus: Seine Stellung in Staat und Gesellschaft*, Mainz 1979 [Akademie der Wissenschaften und der Literatur. Abhandlungen der Geistes- und Sozialwissenschaftlichen Klasse 6].

Kühlborn, J.-B., 'Einige Beobachtungen zum Bau der Stadtmauer der Colonia Ulpia Traiana', *Bonner Jahrbücher* 187 (1987) 475–94.

Laet, S.J. de, *Portorium: étude sur l'organisation douanière chez les Romains, surtout à l'époque du Haut-Empire*, Brussels 1949.

Lafaye, G., 'Plaustrarius', in: C. Daremberg and E. Saglio (eds), *Dictionnaire des antiquités grecques et romaines d'après les textes et les monuments*. Paris 1877–1919, 503–4 [Vol. IV.1].

——, 'Plaustrum', in: C. Daremberg and E. Saglio (eds), *Dictionnaire des antiquités grecques et romaines d'après les textes et les monuments*. Paris 1877–1919, 504–6 [Vol. IV.1].

——, 'Vehiculum', in: C. Daremberg and E. Saglio (eds), *Dictionnaire des antiquités grecques et romaines d'après les textes et les monuments*. Paris 1877–1919, 667–8 [Vol. V].

Lanciani, R., *Forma Vrbis Romae*. Rome 1990.

Laurence, R., *Roman Pompeii: Space and Society*. London and New York 1994/1996.

——, *The Roads of Roman Italy: Mobility and Cultural Change*. London and New York 1999.

Lécrivain, C., 'Statio, stationarius', in: C. Daremberg and E. Saglio (eds), *Dictionnaire des antiquités grecques et romaines d'après les textes et les monuments*. Paris 1877–1919, 1468–9 [Vol. IV.2].

Legras, H., *La table latine d'Héraclée (la prétendue Lex Julia Municipalis)*. Paris 1907.

Lehner, H., 'Ausgrabung- und Fundberichte des Provinzialmuseums in Bonn', *Bonner Jahrbücher* 110 (1903) 122–87 [Regierungsbezirk Düsseldorf 182–7].

Liebenam, W., 'Exercitus', in: A. Pauly *et al.* (eds), *Paulys Realencyclopädie der classischen Altertumswissenschaft*. Stuttgart and Munich 1903–78, 1589–1679 [Vol. VI].

Liebeschuetz, W., 'Rubbish disposal in Greek and Roman cities', in: X. Dupré Raventós and J.-A. Remolà (eds), *Sordes urbis: la eliminación de residuos en la ciudad romana. Actas de la reunión de Roma (15–16 de Noviembre de 1996)*. Rome 2000, 51–61.

Liesen, B., 'Römische Keramik aus dem Bereich des Burginatiumtores der CUT', in: S. Kraus and A. Rieche (eds), *Grabung – Forschung – Präsentation: Sammelband*. Cologne 1994, 237–46 [Xantener Berichte 5].

Lobel, M.D., *Historic Towns: Maps and Plans of Towns and Cities in the British Isles, with Historical Commentaries, from Earliest Times to 1800*. Baltimore 1969–76 [2 vols].

Lorenz, T., *Römische Städte*. Darmstadt 1987.

Lugli, J. and I. Gismondi, *Forma Urbis*. Novara 1949.

MacMullen, R., 'How big was the Roman Imperial army?' *Klio* 62 (1980) 2: 451–60.

Marquardt, J., *Das Privatleben der Römer*. Leipzig 1986[2] [Teil 2: Erwerb und Unterhaltung].

Martha, J., *L'art étrusque*. Paris 1889.

Masquelez and E. Saglio, 'Cataracta', in: C. Daremberg and E. Saglio (eds), *Dictionnaire des antiquités grecques et romaines d'après les textes et les monuments*. Paris 1877–1919, 967 [Vol. I.2].

Matthews Jr., K.D., 'The embattled driver in ancient Rome', *Expedition: The Bulletin of the University Museum of the University of Pennsylvania* 2 (1960) 3: 22–7.

Mau, A., *Pompeii: Its Life and Art.* New York 1899 [transl. F.W. Kelsey].

Meer, L.B. van der, 'L'affresco sotto le Terme di Traiano del Colle Oppio, Roma: realtà e progetto', *Oudheidkundige Mededelingen uit het Rijksmuseum van Oudheden te Leiden* 78 (1998), 63–73.

Meijer, F., 'Reizen in de antieke wereld', *Hermeneus* 73 (2001) 2: 70–86.

Mertens, J., 'Les ponts de la via Traiana dans la traversée du Tavoliere de Foggia', in: L. Quilici and S. Quilici Gigli (eds), *Strade Romane: percorsi e infrastrutture.* Rome 1997, 7–18 [Atlante tematico di topografia antica 2 – 1993].

——, 'Naissance d'une ville, Atuatuca Tungrorum-Tongres', *Revue Archéologique de Picardie* (1984) 3–4: 41–8.

Mertens, J. and W. Vanvinckenroye, *Een Romeins gebouwencomplex extra-muros te Tongeren*, Tongeren 1975 [Publicaties van het provinciaal Gallo-Romeins Museum te Tongeren 22].

Meyer, H., 'Die römischen Alpenstrassen in der Schweiz', *Mitteilungen der Antiquarischen Gesellschaft in Zürich* 13 (1861) Abt. 2, Heft 4: 119–40.

Mollo Mezzena, R., 'La strada romana in Valle d'Aosta', in: L. Quilici and S. Quilici Gigli (eds), *Tecnica stradale Romana.* Rome 1999, 57–72.

Mommsen, T., *Römisches Staatsrecht.* Leipzig 1871–88³ [3 vols].

——, 'Zum römischen Strassenwesen', *Hermes* 12 (1877) 486–91.

Morin, A., 'L'ordre de marche de l'armée romaine: le témoignage de Flavius Josèphe', *Revue des Etudes Anciennes* 104 (2002) 1–2: 145–61.

Morley, N., *Metropolis and Hinterland: The City of Rome and the Italian Economy, 200 B.C.–A.D. 200.* Cambridge 1996.

Naerebout, F.G. and H.W. Singor, *De oudheid: Grieken en Romeinen in de context van de wereldgeschiedenis.* Baarn 1995.

Nap, J.M., 'Dateering en rechtskarakter der z.g. lex Julia municipalis', *Verhandelingen der Koninklijke Akademie van Wetenschappen, afdeeling Letterkunde* 11 (1910) [nieuwe reeks].

Nelis-Clément, J., *Les beneficiarii: militaires et administrateurs au service de l'empire (Ier s. a.C. – VIe s. p.C.).* Paris 2000.

Nissen, H., *Pompeianische Studien zur Städtekunde des Altertums.* Leipzig 1877.

Nollé, J., 'Stadtsanierung in der Antike', *Die Alte Stadt* 22 (1995) 1: 30–50.

Ortalli, J., 'La Cispadana orientale: Via Emilia e altre strade', in: L. Quilici and S. Quilici Gigli (eds), *Tecnica stradale Romana.* Rome 1999, 147–60 [Atlante tematico di topografia antica 1].

Overbeck, J., *Pompeji in seinen Gebäuden: Alterthümern und Kunstwerken.* Leipzig 1884 [Vierte im vereine mit August Mau].

Pace, B., *I mosaici di Piazza Armerina.* Rome 1955.

Palmer, R.E.A., 'Customs on market goods imported into the City of Rome', in: J.H. D'Arms and E.C. Kopff (eds), *The Seaborne Commerce of Ancient Rome: Studies in Archaeology and History.* Rome 1980, 217–33.

——, *The Archaic Community of the Romans.* Cambridge 1970.

Peacock, D.P.S. and V.A. Maxfield, *Mons Claudianus: Survey and Excavation 1987–1993.* Cairo 1997 [Institut Français d'Archéologie Orientale 37].

Pekáry, T., *Untersuchungen zu den römischen Reichsstrassen.* Bonn 1968.

Petrikovits, H. von, *Die Rheinlande in römischer Zeit: mit einem Überblick über die rheinische Urgeschichte.* Düsseldorf 1980 [I. Textteil; II. Bild und Dokumentarteil].

Platner, S.B. and T. Ashby, *A Topographical Dictionary of Ancient Rome*. Oxford and London 1929.

Pöhlmann, R., *Die Übervölkerung der antiken Grosstädte im Zusammenhange mit der Gesammtentwicklung städtischer Civilisation*. Leipzig 1884 [Preisschriften gekrönt und herausgegeben von der Fürstlich Jablonowski'schen Gesellschaft zu Leipzig 16, Historisch-nationalökonomische Section 24].

Pouilloux, J., *Choix d'inscriptions grecques: textes, traductions et notes*. Paris 1960.

Quilici, L, 'Evoluzione della tecnica stradale nell'Italia centrale', in: L. Quilici and S. Quilici Gigli (eds), *Tecnica stradale Romana*. Rome 1999, 19–32 [Atlante tematico di topografia antica 1].

——, 'La via Salaria da Roma all'alto Velino: la tecnica strutiva dei manufatti stradali', in: L. Quilici and S. Quilici Gigli (eds), *Strade Romane: percorsi e infrastrutture*. Rome 1997, 85–154 [Atlante tematico di topografia antica 2 – 1993].

Quilici Gigli, S., 'La via Nomentana da Roma ad Eretum', in: L. Quilici and S. Quilici Gigli (eds), *Strade Romane: percorsi e infrastrutture*. Rome 1997, 45–83 [Atlante tematico di topografia antica 2 – 1993].

Radke, G., 'Viae publicae', in: A.F. von Pauly, K. Ziegler and W. Sontheimer (eds), *Der kleine Pauly: Lexikon der Antike*. Stuttgart 1964–75, 1243–6 [Vol. V].

——, 'Viae publicae Romanae', in: A. Pauly *et al.* (eds), *Paulys Realencyclopädie der classischen Altertumswissenschaft*. Stuttgart and Munich 1903–78, 1417–1686 [Suppl. 13].

Raepsaet, G., 'Landtransport', in: H. Cancik and H. Schneider (eds), *Der neue Pauly: Enzyklopaedie der Antike*. Stuttgart 1996–, 1097–1106 [Vol. VI].

——, 'Transport de tambours de colonnes du Pentélique à Eleusis au IVe siècle avant notre ère', *L'Antiquité Classique* 53 (1984) 101–36.

Rathmann, M., '*Viae publicae* in den Provinzen des Imperium Romanum: Probleme einer rechtlichen Definition', in: E. Olshausen and H. Sonnabend, *Stuttgarter Kolloquium zur Historischen Geographie des Altertums 7, 1999: Zu Wasser und zu Land: Verkehrswege in der antiken Welt*. Stuttgart 2002, 410–18 [Geographica Historica 17].

Reinach, S., 'Elephas', in: C. Daremberg and E. Saglio (eds), *Dictionnaire des antiquités grecques et romaines d'après les textes et les monuments*. Paris 1877–1919, 536-44 [Vol. II.1].

Rieche, A., *Gids door het Archeologisch Park Xanten*. Cologne 1994.

Riedel, M., *Köln, ein römisches Wirtschaftszentrum*. Cologne 1982.

Riepl, W., *Das Nachrichtenwesen des Altertums mit besonderer Rücksicht auf die Römer*. Leipzig and Berlin 1913.

Rogge, M., 'Het wegennet van de Romeinen', *Hermeneus* 73 (2001) 2: 102–14.

Röring, C.H., *Untersuchungen zu römischen Reisewagen*. Koblenz 1983.

Rottier, H., *Stedelijke structuren: een inleiding tot de ontwikkeling van de Europese stad*, Muiderberg 1980.

Royo, M., 'Le palais dans la ville', *Mélanges de l'Ecole Française de Rome. Antiquité* 106 (1994) 1: 219–45.

Säflund, G., *Le mura di Roma repubblicana: Saggio di archeologica romana*. Uppsala 1932.

Saglio, E., 'Carpentum', in: C. Daremberg and E. Saglio (eds), *Dictionnaire des antiquités grecques et romaines d'après les textes et les monuments*. Paris 1877–1919, 926–7 [Vol. I.2].

——, 'Carruca', in: C. Daremberg and E. Saglio (eds), *Dictionnaire des antiquités grecques et romaines d'après les textes et les monuments*. Paris 1877–1919, 928 [Vol. I.2].

——, 'Carrus', in: C. Daremberg and E. Saglio (eds), *Dictionnaire des antiquités grecques et romaines d'après les textes et les monuments*. Paris 1877–1919, 928–9 [Vol. I.2].

——, 'Cisium', in: C. Daremberg and E. Saglio (eds), *Dictionnaire des antiquités grecques et romaines d'après les textes et les monuments*. Paris 1877–1919, 1201 [Vol. I.2].

——, 'Rheda', in: C. Daremberg and E. Saglio (eds), *Dictionnaire des antiquités grecques et romaines d'après les textes et les monuments*. Paris 1877–1919, 862–3 [Vol. IV.2].

Sakai, S., 'Il problema dell'esistenza della cosiddetta Porta Capua a Pompei', *Opuscula Pompeiana* 12 (2003–4) 27–63.

Santillo Frizell, B., 'Sheep, shrine, sulphur: Via Tiburtina and the urban pastoral landscape', in: H. Bjur and B. Santillo Frizell (eds), *Via Tiburtina: Space, Movement and Artefacts in the Urban Landscape*. Rome 2005, 31–7 [The Swedish Institute in Rome. Projects and Seminars 4:1].

Sarton, G., *A History of Science*. Cambridge, MA 1952–9 [2 vols].

Sayar, M.H., 'Antike Straßenverbindungen Kilikiens in der römischen Kaiserzeit', in: E. Olshausen and H. Sonnabend, *Stuttgarter Kolloquium zur Historischen Geographie des Altertums 7, 1999: Zu Wasser und zu Land: Verkehrswege in der antiken Welt*. Stuttgart 2002, 454–73 [Geographica Historica 17].

Schmidt, B., 'Ein dendrochronologischer Befund zum Bau der Stadtmauer der Colonia Ulpia Traiana', *Bonner Jahrbücher* 187 (1987) 495–503.

Schneider, G., 'Alle Wege führen nach Rom – auch Geleisestraßen?' in: H. Koschik (ed.), *'Alle Wege führen nach Rom . . .': Internationales Römerstraßenkolloquium Bonn*. Pulheim Brauweiler 2004, 67–78 [Materialien zur Bodendenkmalpflege im Rheinland 16].

Schneider, H.-C., *Altstraßenforschung*. Darmstadt 1982.

Schultze, R., 'Die römischen Stadttore', *Bonner Jahrbücher* 118 (1909) 280–352.

Schulzki, H.J., 'Leuga', in: H. Cancik and H. Schneider (eds), *Der neue Pauly: Enzyklopaedie der Antike*. Stuttgart 1996–, 99–100 [Vol. VII].

Scobie, A., 'Slums, sanitation, and mortality in the Roman world', *Klio* 68 (1986) 2: 399–433.

Scullard, H.H., *Kleine atlas van de antieke wereld: editie voor het onderwijs*. Amsterdam and Brussels 1964^2.

——, *The Elephant in the Greek and Roman World*. Ithaca, NY 1974.

——, *The Etruscan Cities and Rome*. London and Southampton 1967.

Simonelli, A., 'Considerazioni sull'origine, la natura e l'evoluzione del *pomerium*', *Aevum* 75 (2001) 119–62.

Singor, H.W., 'Historisch overzicht: Constantinopel tot het begin van de 7de eeuw', *Hermeneus* 68 (1996) 2: 50–61.

Sizoo, A., *Reizen en trekken in de Oudheid*. Kampen 1962.

Smeesters, J., *De Romeinse monumenten van Tongeren*. Tongeren 1975.

Sonnabend, J., 'Stadtverkehr im antiken Rom: Probleme und Lösungsversuche', *Die Alte Stadt* 19 (1992) 3: 183–94.

Staccioli, R.A., *The Roads of the Romans*. Los Angeles 2003 [transl. of *Strade dei Romani*. Milan 2003].

Stambaugh, J.E., *The Ancient Roman City: Ancient Society and History*. Baltimore and London 1988.

Stark, R., *The Rise of Christianity: A Sociologist Reconsiders History*. Princeton, NJ 1996.

Stephan, H. von, *Das Verkehrsleben im Altertum und im Mittelalter*. Goslar-Braunschweig 1966^2.

Stoffel, P., *Über die Staatspost, die Ochsengespanne und die requirierten Ochsengespanne: eine Darstellung des römischen Postwesens auf Grund der Gesetze des Codex Theodosianus und des Codex Iustinianus*. Bern 1994 [Reihe III Geschichte und ihre Hilfswissenschaften; Vol. 595].

Stuart, P. and M.E.T. de Grooth (eds), *Langs de weg*. Heerlen and Maastricht 1987.
Süßenbach, U., *Die Stadtmauer des römischen Köln*. Cologne 1981.
Teitler, H.C., 'De Romeinse postdienst', *Hermeneus* 73 (2001) 2: 130–5.
Thesaurus Linguae Latinae, Leipzig 1900–.
Thomas, R., 'Gräber und Wohnbauten vor der westlichen römischen Stadtmauer in Köln', *Kölner Jahrbuch für Vor- und Frühgeschichte* 23 (1990) 401–12.
Thulin, C.O., *Die etruskischen Disciplin*. Göteborg 1906–9 [3 vols].
Tilburg, C.R. van, *Romeins verkeer: weggebruik en verkeersdrukte in het Romeinse Rijk*. Amsterdam 2005 [Serie De Oudheid].
——, 'Stadspoorten in het Romeinse Rijk', *Spiegel Historiael* 39 (2004) 9: 391–6.
——, 'Verkeer in het Romeinse Rijk', *Spiegel Historiael* 37 (2002) 5: 192–7.
Tsujimura, S., 'Ruts in Pompeii: the traffic system in the Roman city', *Opuscula Pompeiana* 1 (1991) 58–90.
Valenti, M., 'Esempi di infrastrutture stradali nel Cassinate', in: L. Quilici and S. Quilici Gigli (eds), *Strade Romane: percorsi e infrastrutture*. Rome 1997, 35–44 [Atlante tematico di topografia antica 2 – 1993].
Vanvinckenroye, W., *Tongeren Romeinse stad*. Tongeren 1975 [Publicaties van het provinciaal Gallo-Romeins Museum te Tongeren 23].
Varène, P., *L'enceinte gallo-romaine de Nîmes: les murs et les tours*. Paris 1992 [53e supplément à 'Gallia'].
Virlouvet, C., 'La topographie des distributions frumentaires avant la création de la *Porticus Minucia Frumentaria*', in: Centre National de la Recherche Scientifique Paris and l'Ecole Française de Rome (eds), *L'urbs: espace urbain et histoire (Ier siècle av. J.-C. – IIIe siècle ap. J.-C. Actes du colloque international organisé par le Centre National de la Recherche Scientifique et l'Ecole Française de Rome (Rome, 8–12 mai 1985)*, Rome 1987, 175–89 [Collection de l'Ecole Française de Rome 98].
Vliet, M. van, *Zwaar transport over land in de Griekse en Romeinse oudheid* [unpublished MA thesis].
Wacher, J., 'Second-century cities in Britannia', in: H.-J. Schalles, H. von Hesberg and P. Zanker (eds), *Die römische Stadt im 2. Jahrhundert n. Chr.: Der Funktionswandel des öffentlichen Raumes*, Cologne 1992, 13–31 [Xantener Berichte 2].
——, *The Towns of Roman Britain*. London and New York 1997[2] [paperback].
Waldherr, G., 'Das System der *calles* (Herdenwanderwege) im römischen Italien: Entstehung und infrastrukturelle Bedeutung', in: E. Olshausen and H. Sonnabend, *Stuttgarter Kolloquium zur Historischen Geographie des Altertums 7, 1999: zu Wasser und zu Land: Verkehrswege in der antiken Welt*. Stuttgart 2002, 429–44 [Geographica Historica 17].
Wallace-Hadrill, A., 'Public honour and private shame: the urban texture of Pompeii', in: T.J. Cornell and K. Lomas (eds), *Urban Society in Roman Italy*. London 1995, 39–62.
Walser, G., *Via per Alpes Graias: Beiträge zur Geschichte des kleinen St. Bernhard-Passes in römischer Zeit*. Stuttgart 1986 [Historia 48].
Ward-Perkins, J.B., *Cities of Ancient Greece and Italy: Planning in Classical Antiquity*. New York 1974.
Weber, W., 'Das Ehrenrecht des Wagenfahrens in römischen Städten', in: H. Beck and P.C. Bol (eds), *Spätantike und frühes Christentum*, Frankfurt 1983, 208–11.
Webster, G., *The Roman Imperial Army of the First and Second Centuries A.D.* Norman, OK 1998[3].
Wegener Sleeswyk, A., *Wielen, wagens, koetsen*. Leeuwarden 1992.
White, K.D., *Greek and Roman Technology*. London 1984.

Wiesehöfer, J., 'Bevölkerung, Bevölkerungsgeschichte', in: H. Cancik and H. Schneider (eds), *Der neue Pauly: Enzyklopaedie der Antike.* Stuttgart 1996–, 599–605 [Vol. II].

Willems, W.J.H., *Romeins Nijmegen: vier eeuwen stad en centrum aan de Waal.* Utrecht 1990.

Wit, F. de, *Theatrum ichnographicum omnium urbium et praecipuorum oppidorum belgicarum XVII provinciarum peraccurate delineatarum = Perfecte aftekeningen der steden van de XVII Nederlandse provinciën in plattegronden = Le théâtre des plans de toutes les villes qui sont situéz dans les XVII provinces du Pays Bas parfaictement deseignéz.* Amsterdam 1980 [Facsimile reprint, original edition 1698].

Yeo, C.A., 'Land and sea transportation in imperial Italy', *Transactions and Proceedings of the American Philological Association* 77 (1946) 221–44.

Zimmermann, K., '"Verkehrsregelungen" in der Antike', in: E. Olshausen and H. Sonnabend, *Stuttgarter Kolloquium zur Historischen Geographie des Altertums 7, 1999: zu Wasser und zu Land: Verkehrswege in der antiken Welt.* Stuttgart 2002, 181–201 [Geographica Historica 17].

INDEX LOCORUM

INDEX GENERALIS

Related titles from Routledge

Britannia
The Creation of a Roman Province
John Creighton

In the Late Iron Age two kings held dominion over much of Lowland Britain: Cunobelin and Verica. Just before AD 43 the rule of both of them ended – one died and the other fled – and Rome, under the Emperor Claudius, took the opportunity to invade.

Within a few generations the ceremonial centres of these Iron Age kings had been transformed into the magnificence of Roman towns with their monumental public buildings. This book looks at the interface of these two worlds, Iron Age and Roman, to see how much each owed the other.

Britannia: the creation of a Roman province offers a complete re-evaluation of both the evidence for and the interpretation of the rule of the kings of Late Iron Age Britain on the eve of the Roman conquest, and their long-lasting legacy in the creation of the Roman province of Britannia. Among topics considered are:

- the links between Iron Age kings of Britain and Rome before the Claudian conquest
- the creation of the towns of Roman Britain
- the many different natures of 'Roman Identity'
- The long lasting influence of the kings on the development of the province
- the widely different ways that archaeologists have read the evidence

ISBN10: 0-415-33313-X (hbk)
ISBN13: 978-0-415-33313-9 (hbk)

Related titles from Routledge

The Romans
An Introduction
Antony Kamm

The Romans provides students with an excellent introduction to all aspects of Roman culture. Antony Kamm focuses on literature, art and architecture, and includes discussion of major contemporary developments in the study of ancient Rome.

All the crucial elements of Roman history such as the reign of the Caesars and the role of the army are discussed. Again drawing on recent scholarship, the author examines important themes like imperialism, religion and everyday social life.

Incorporating maps, charts and pictures as well as a comprehensive bibliography and index, *The Romans* is a concise yet thorough introductory survey of Roman civilization. An indispensable textbook, *The Romans* will prove invaluable for students of disciplines such as ancient history, art, literature or history, who require a reliable, accessible guide to all aspects of Roman culture.

ISBN10: 0-415-12039-X (hbk)
ISBN10: 0-415-12040-3 (pbk)

ISBN13: 978-0-415-12039-5 (hbk)
ISBN13: 978-0-415-12040-1 (pbk)

Related titles from Routledge

War and Society in Imperial Rome, c.31BC to AD 280
Brian Campbell

This well-documented study of the Roman army provides a crucial aid to understanding the Roman Empire in economic, social and political terms. Employing numerous examples, Brian Campbell explores the development of the Roman army and the expansion of the Roman Empire from 31 BC-280 AD.

Dealing with issues such as the financial implications of supporting a professional army in war and peace, Brian Campbell explores the wider significance of the army and warfare in Roman life and culture. This superbly researched survey provides readers with an invaluable guide to this important and much neglected subject.

ISBN10: 0-415-27881-3 (hbk)
ISBN10: 0-415-27882-1 (pbk)

ISBN13: 978-0-415-27881-2 (hbk)
ISBN13: 978-0-415-27882-9 (pbk)

For Product Safety Concerns and Information please contact our EU representative GPSR@taylorandfrancis.com
Taylor & Francis Verlag GmbH, Kaufingerstraße 24, 80331 München, Germany

www.ingramcontent.com/pod-product-compliance
Lightning Source LLC
LaVergne TN
LVHW011122300926
849160LV00008B/484
9780415512619